T3-BOC-439

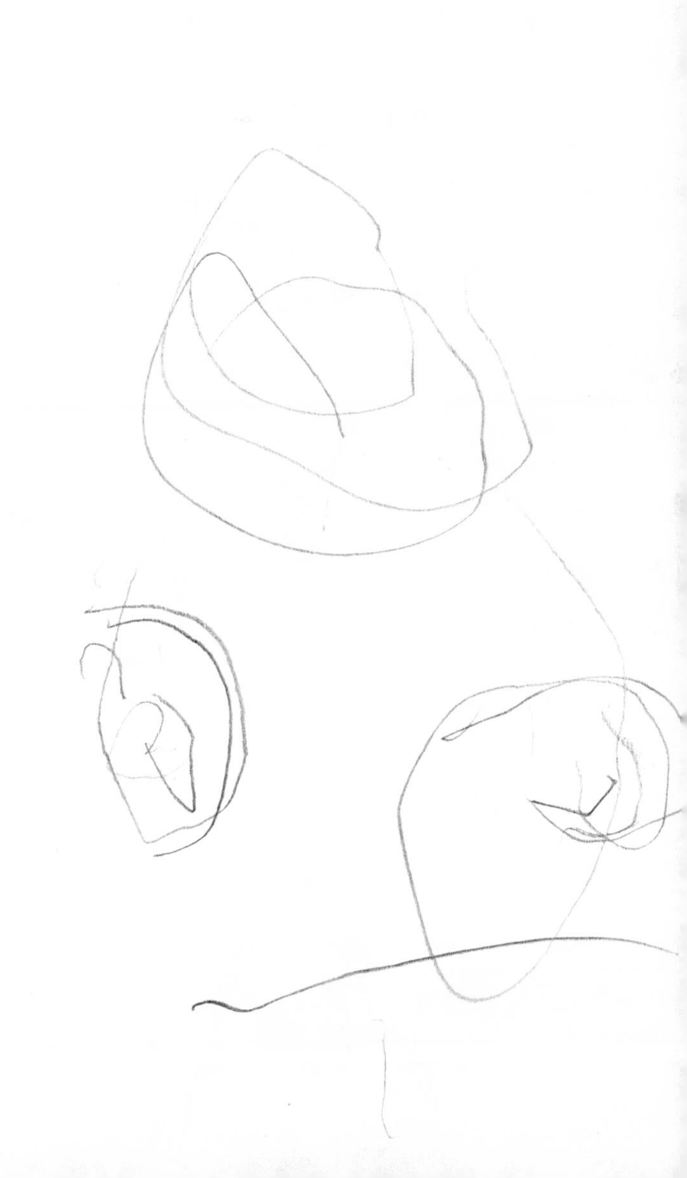

AN INTRODUCTION TO PROGRAMMING IN LISP

AN INTRODUCTION TO PROGRAMMING IN LISP

H. Wertz

Université Paris VIII & CNRS

JOHN WILEY & SONS
Chichester · New York · Brisbane · Toronto · Singapore

Library of Congress Cataloging-in-Publication Data

Wertz, H. (Harald), 1947–
An introduction to programming in LISP.

Translation of: LISP, une introduction à la programmation.
Bibliography: p.
Includes index.
1. LISP (Computer program language) I. Title.
QA76.73.L23W4713 1988 005.13′3 87–20973
ISBN 0 471 91490 8 (pbk.)

British Library Cataloguing in Publication Data

Wertz, H.
An introduction to programming in LISP.
1. LISP (Computer program language)
I. Title
005.13′3 QA76.73.L23

ISBN 0 471 91490 8

Typeset by Photo·Graphics Ltd, Honiton, Devon.
Printed and bound in Great Britain by Anchor Brendon Ltd, Tiptree, Essex.

CONTENTS

FOREWORD

LISP is one of the oldest programming languages: its first implementations appeared at the end of the 1960s, only a few years after FORTRAN.

LISP was designed by John McCarthy to process symbolic expressions. From the outset, it was used to write programs for differential and integral symbolic calculus, for electronic circuit theory, for mathematical logic and for games programming.

LISP is also one of the most widely used programming languages: it runs on virtually any computer, of any size, from any manufacturer.

LISP is one of the liveliest programming languages. The astonishing number of different versions, such as MACLISP, INTERLISP, Common-LISP, VLISP, Le_LISP, etc., bear witness to its vivacity. Furthermore, machines have appeared using an architecture specially designed for the implementation of LISP, e.g. those manufactured by Symbolics or by the LISP-Machine Company, or the Maia machine, currently in use at the CGE.

In LISP, as in machine languages, data representation and program representation are identical: a LISP program may therefore build other LISP programs or modify itself in the course of execution. Above all, this identity of representation makes it possible to write LISP in LISP itself.

LISP is an interactive language with an integrated programming environment. This not only implies that all programs and all data are accessible, modifiable and analysable in LISP itself, without moving out of the language, but also that the construction of programs is considerably accelerated: the cycle *text editing – compiling – link editing – loading – executing* of classical compiled languages is reduced to the cycle *writing – evaluating*.

These advantages, along with the syntactic simplicity of LISP, the possibility of writing an algorithm immediately without passing through the variable or type declaration stage, and finally the fact that LISP is based on recursion, make LISP an excellent language for learning or teaching programming.

Learning LISP is the theme of this introduction to programming. It makes no assumptions concerning the reader's prior knowledge. It uses commented examples to introduce the LISP language, its data structures, its control structures and its programming techniques.

As we suggested above, LISP appears in various versions: despite regular efforts to establish a standard LISP, the development of this magnificent language has entailed a large variety of *dialects*. Until recently, LISP was primarily confined to university and research environments. It took the commercial success of expert and artificial intelligence systems to exert enough pressure on the LISP community to develop a standardized kernel of LISP. This kernel language is called Common-LISP. Throughout this book we will use two versions of LISP: CCL, an implementation of Common-LISP developed at the University of Paris VIII (Vincennes) by Patrick Greussay, and Le_LISP, a dialect derived from VLISP and Common-LISP developed by Jerome Chailloux at INRIA. Thus, the book handles the two most common versions of LISP in Europe. Where differences occur between these two dialects, they will be pointed out.

Each chapter is made up of two parts: the first is a description of the new concepts tackled in the chapter, the second a series of exercises for which solutions are given at the end of the book. These exercises are an integral part of each chapter. It is often assumed in the text that the exercises for the previous chapter have been understood.

The book's structure is as follows:

- The first three chapters introduce the standard data structures in LISP, i.e. atoms and lists, as well as the basic functions allowing them to be accessed and constructed. Without an understanding of these structures and functions, we cannot write any programs.
- In Chapter 4 we study the possibilities open to users to define their own functions. This is the basis of all LISP programming. We shall also examine the question of binding of variables.
- In Chapter 5 we shall consider the most common predicates as well as the decision function IF.
- Chapter 6 uses four examples to introduce the concept of recursion, the most common form of repetition in LISP.
- Chapter 7 considers arithmetic functions and the basis for numeric programming in LISP.
- In Chapters 8 and 9 we shall return to the concept of atoms and introduce the use of P-lists to associate multiple values with an atom. We shall also consider a particular application of P-lists: functions that store a trace of their own activity.
- Chapter 10 deals with printing.
- Chapters 11 and 12 show different ways of defining functions. We shall also consider the functions EVAL and APPLY.
- In Chapter 13 we shall examine input/output functions as well as the role of various special characters in LISP.
- In Chapter 14 we shall build up a small pattern matching function.
- In Chapters 15 and 16 we shall examine the structure of lists in detail. We shall also consider assignment functions and a third type of function: macro-functions.
- Chapter 17 describes various forms of iterative repetition.
- In Chapter 18 we shall return to the example of a pattern matching function. We

shall build a very powerful pattern matcher, incorporating all the concepts discussed before.
- Finally, Chapter 19 gives the solutions to all the exercises in the book.

This study will not cover all the functions available in LISP: only the most important will be discussed. However, the subset of functions examined is more than large enough to write any program in any system. Nor shall we consider the implementation of LISP, its compilation and – above all – its application in artificial intelligence. These questions will be dealt with in a second volume. However, the bibliography at the end of the book also includes works on these other aspects of LISP.

The contents of this book are based on an introductory course to high level languages that I have been giving for some years at the University of Paris VIII (Vincennes). I should like to thank the students who, through their critical remarks, their questions and their enthusiasm, exerted a major influence on the writing of this book. The book is dedicated to all of them.

The book was prepared by the author at IRCAM on a DEC VAX–780 computer under the UNIX system, thanks to the generosity of Pierre Boulez and David Wessel. It would not have been possible without the support of the IRCAM technical team, in particular Michèle Dell-Prane and Patrick Sinz.

This book has, of course, benefited from suggestions, criticisms and comments made by many people. I should like in particular to thank Jean-Pierre Briot, Annette Cattenat, Frederic Chauveau, Pierre Cointe, Charles Consel, Gerard Dahan, Yves Devillers, Jacques Ferber, Patrick Greussay, Daniel Goossens, Eric Halle, Christian Jullien, Jean Mehat, Gerard Nowak, Gerard Paul, Yves-Daniel Perolat, Jean-François Perrot, Christian Riesner, Nicole Rocland, Bernard-Paul Serpette, Jacqueline Signorini, Patrick Sinz, Ellen Ann Sparer and Roger Tanguy.

1 PRELIMINARIES

The language used in this introduction to programming is LISP.
LISP is an acronym for List Processor.
As its name implies, LISP handles LISTS. A list is something which begins with an opening parenthesis '(' and ends with a closing parenthesis ')'. It is that simple. Here are a few examples of lists:

```
(DO RE MI)
(OH LA LA)
(THE CUBE IS ON THE TABLE)
```

Syntactically a list is defined as:

list ::= ([x1 x2 . . . xn])

and x is defined as:

x ::= list | atom

The sign :: = means *is defined as* and the sign | means *or*. Anything between curly brackets, { and }, is optional. This means that it can be present or not, depending on the circumstances. The definition above may therefore be read as 'x is defined as a list or an atom'.

Let us go on with our definitions:

atom	:: =	number | name
number	::=	0 | 1 | 2 | 3 | . . . | 1024 | . . . | -1 | -2 | . . . | -1024 | . . . (which means that it can be a positive or negative number)
name	::=	a sequence of alphanumeric characters containing at least one letter and no *separators*
separator	::=	. | space | (|) | [|] | ' | tab | carriage return which are characters with a particular role.

A few examples of atoms:

```
DO
RE
```

```
MI
THIS-IS-A-VERY-LONG-ATOM
CUBE
CUBE1
1A
```

Here are some examples of LISP objects:

```
128              ;a number;
-32600           ;a negative number;
HAHA             ;a name atom (alphanumeric atom);
()               ;a list with 0 elements;
(HAHA)           ;a list with 1 element;
(A LIST)         ;a list with 2 elements;
((A) LIST)       ;a list with 2 elements, the first;
                 ;a list with 1 element,;
                 ;the second element is an atom;
(A (LIST))       ;another list with 2 elements;
```

Here are three examples of things which cannot be objects in LISP:

```
)                    ;nothing can begin with a closing bracket;
(WELL (WELL (WELL    ;here a lot of closing brackets are missing;
(. . .)              ;unfortunately the '.' character is not allowed in LISP (yet).
```

The elements of a list may be counted, so we can talk of the *length* of a list. The *length* of a list is the number of its elements.

Here are some lists of length 2:

```
(1 2)
(WELL WELL)
( ((AHA)) (YET ANOTHER LIST) )
(()())
```

and here are a few other lists with their lengths:

lists	length
()	0
(JOHN GIVES MARY A BOOK)	5
((SUBJECT) (VERB) (OBJECT))	3
(((X + Y) +Z) → (X+ (Y+Z)))	3
(SOMEBODY LOVES MARY)	3
(WHAT THE HELL ARCHY)	4
(1 2 3 A B C C B A 3 2 1)	12

The empty list, (), is called NIL. In LISP the NIL atom and the list () have the same value, i.e. (). You can write it either in the form of an atom: NIL or in the form of a list: ().

Clearly, parentheses are always balanced: a list, and therefore any sublist, always has the same number of opening and closing parentheses. Moreover, when reading a list character by character, from left to right, the number of closing parentheses met is always less than the number of opening parentheses, except, of course, at the end of the list where the number of opening parentheses is equal to the number of closing parentheses.

If you have problems determining whether an object represents a well-formed list, here is a little algorithm which can help you:

> If the object in question does not begin with an opening parenthesis, then quite clearly you are not dealing with a list and the question is answered. Otherwise, count parentheses as follows: imagine that you have a counter which is initially set to 0. Then read the object from left to right. Each time you meet an opening parenthesis increase the value of your counter by one and mark the parenthesis with the current value of your counter. If you meet a closing parenthesis, mark it with the current value of the counter, and then decrease the value of that counter by 1. If, before reaching the end of the list, the counter at any moment reaches a number less than or equal to 0, then the object is not a well-formed list. On the other hand, if you reach the end of your object with your counter at 0, then the object is a perfectly correct list.

Here are a few example applications of this algorithm:

- The object (((AHA)) (ANOTHER LIST)) may be labelled as follows:

  $$(_1(_2(_3\text{AHA})_3)_2)\ (_2\text{ANOTHER LIST})_2)_1$$

  The final parenthesis is labelled '1': the object is therefore indeed a list.

- Let us test the following object: (HEY? (WHAT?)))

  $$(_1\text{HEY? }(_2\text{WHAT?})_2)_1)_0$$

  Here there is one closing parenthesis too many.

- Finally: (((X + Y) → (X +(Y + Z))) gives

  $$(_1(_2(_3X + Y)_3 + Z) \rightarrow (_2X + (_3Y + Z)_3)_2)_1$$

  Well then? is this a list?

This way of numbering parentheses also allows easy identification of the different elements of a list, the different elements of sub-lists, of sub-sub-lists, etc. Take, for example, the last list:

$$(_1(_2(_3X + Y)_3 + Z)_2 \rightarrow (_2X + (_3Y + Z)_3)_2)_1$$

This is a list with three elements: the two sub-lists enclosed by the parentheses labelled '2' and the atom '→'. Or, more precisely, the parentheses are numbered in a way that corresponds to the *depth* of the list: the list (X + Y) is at depth 2, the list (X + (Y + Z)) is at depth 1, etc. A list enclosed by parentheses numbered *x* is at depth *x-1*, and all the elements within that list are at depth *x*.

1.1 EXERCISES

1. Which of the objects below are lists, atoms or unacceptable objects in LISP?

```
123
(EIN (SCHONES) BUCH)
(AHA (ASTONISHING...!))
(((((1)2)3)4)5)
-3Aiii
T
(((((ARRRGH)))))
```

2. How many elements are there in the following lists:

```
(EIN(SCHONES)BUCH)
(((((1)2)3)4)5)
(((((ARRRGH)))))
(UNIX (IS A) TRADEMARK (OF) BELL LABS)
```

3. For each of the lists above, give the element of greatest depth and state its depth.

2 BASIC FUNCTIONS: QUOTE, CAR, CDR, CONS

The objects with which you will be working in LISP are therefore lists and atoms. There are special lists, called *forms*, which tell the LISP machine that it must do something. In LISP, 'doing something' is referred to as 'calling a function'. Let us start by defining a form:

form ::= (*function-name* [argument$_1$ argument$_2$ . . . argument$_n$])

A *function call* or an *evaluation* of a form *returns* a value. Any LISP function returns a value.

Here is the definition of a *very* important function, QUOTE:

(QUOTE arg_1) → arg_1

This tells the machine on what object it must work; the *value it returns* is the LISP object given as argument (to the function) itself. *Returns as value* will be written '→'.

A few examples of calls to function QUOTE are:

```
(QUOTE WELL)                          → WELL
(QUOTE (A B C)                        → (A B C)
(QUOTE (THE (PUMPKIN(EATER))))        → (THE (PUMPKIN (EATER)))
```

The QUOTE function is extremely useful in LISP programming: since most LISP functions evaluate their arguments, we use the QUOTE function for arguments that we do not want to evaluate. Evaluation of the arguments of functions can therefore be allowed to take place without qualms: QUOTE will return its argument *as it stands*.

The QUOTE function is so important and so often used that there is an abbreviated notation for it:

```
'WELL                                 → WELL
'(A B C)                              → (A B C)
'(THE (PUMPKIN (EATER)))              → (THE (PUMPKIN (EATER)))
```

This notation is merely an abbreviation: the machine understands the same thing by it, i.e. it is a call to function QUOTE. The QUOTE function can be thought of as the identity function.

As we are dealing with lists, and as the elements of a list may be enumerated, there are functions to access the different elements of a list. First of all there is the function CAR:[1]

(CAR *arg*) → the first element of list *arg* given as argument. *Arg* must be a list.

Here are a few examples of calls:

```
(CAR '(A B C)                        → A
```

note that the argument was QUOTED. Why?

```
(CAR '(THE (PUMPKIN) EATER)))        → THE
(CAR '(((O C N H)) P S))             → ((O C N H))
```

Naturally, LISP functions may be combined:

```
(CAR (CAR '(((O C N H))P S)))        → (O C N H)
(CAR (CAR (CAR '(((O C N H))P S))))  → O
```

(footnote [2] attached to (O C N H))

Moreover:

```
(CAR '(CAR '(((O C N H))P S)))       → CAR
```

Do you now see the use of the QUOTE function?

The function CDR may be defined as follows:

(CDR *arg*) → list *arg* is given as argument without the first element. *arg* must be a list.

CDR is the complementary function of CAR. Here are a few examples:

```
(CDR '(A B C))                       → (B C)
(CDR '(THE (PUMPKIN (EATER))))       → ((PUMPKIN (EATER)))
(CDR '(((O C N H) P S))              → (P S)
```

---

[1] This peculiar term has only been retained for historical reasons: the first implementation of LISP was on an IBM-704 machine. A word on this machine was divided into an 'address' part and a 'decrement' part. The Contents of the Address of Register gave the CAR of a list and the Contents of the Decrement of Register gave the CDR of a list. In honour of this first implementation of LISP, all later versions maintained these two names.

[2] Yes: O, C, N, H, P and S are the chemical elements necessary for the emergence of life.

and combining these functions:

```
(CDR(CDR '(A B C)))                    → (C)
(CAR (CDR '(A B C)))                   → B
(CAR (CDR (CDR '(A B C))))             → C
```

Certain combinations of functions CAR and CDR are so useful that there are abbreviated notations for them:

(CAR (CDR *list*)) is the same thing as (CADR *list*)
(CAR (CDR (CDR *list*))) is the same thing as (CADDR *list*)

The function CADR returns the second element of the argument *list*, and the function CADDR returns the third element.

Check for yourself that with appropriate combinations of functions CAR and CDR, you can obtain *any* element in *any* list.

So far we have only met a function returning its argument as it stands (QUOTE), a function returning the first element of a list (CAR) and a function (CDR) which returns the rest of a list, i.e. a list without its first element. We also need a function to CONStruct a new list, which is the function CONS, defined as:

(CONS *argument list*) → list *list* with the value of *argument 1* as the new first element.

e.g.:

```
(CONS 'A '(B C))                       → (A B C)
(CONS '(A B) '((CD))                   → ((A B) C D)
(CONS (CAR '(A B C))(CDR'(A B C)))     → (A B C)
(CAR (CONS 'A '( B C)))                → A
(CDR (CONS 'A '(B C)))                 → (B C)
```

To end this first part, here is the representation of a brief interaction with the machine (the machine displays a '?' when it expects you to enter something; it displays the value of what you have asked, preceded by the sign '=' on the next line):[3]

```
? 'A
= A

? '(A B C)
= (A B C)
```

[3] Note that the 'prompt character' varies for different implementations. Often you may find '→' instead of the question-mark and the '=' may be omitted.

```
? '''(A B C)
= ''(A B C)

? (CAR '(A B C)
= A

? (CDR '(A B C))
= (B C)

? (CADR '(A B C)
= B

? (CADDR '(A B C))
= C

? (CONS 'A '(B C))
= (A B C)

? (CONS 'A (CONS 'B'()))
= (A B)

? (CONS (CAR '(A B C)) (CDR '(A B C)))
= (A B C)
```

Important note:

By definition of the functions, both the CAR and CDR of NIL are equal to NIL. We therefore have the following relations:

```
(CAR NIL)    → ()
(CAR ())     → ()
(CDR NIL)    → ()
(CDR ())     → ()
```

Note too:

```
(CONS () '(A B C)) = (CONS NIL '(A B C)) → (() A B C)
```

Clearly, to carry out a CONS with a NIL and a list means inserting an empty list at the head of the list given as second argument. NIL as a second argument of a CONS corresponds to putting a pair of parentheses around the first argument of CONS:

```
(CONS 'A NIL) = (CONS 'A ()) → (A)
```

2.1 EXERCISES

1. Give the results of the following function calls:
   a. (CAR '((A(B C)) D (E F))) → ?
   b. (CDR '((A (B C)) D (E F))) → ?
   c. (CADR (CAR '((A (B C)) D (E F)))) → ?
   d. (CADDR '((A (B C)) D (E F))) → ?
   e. (CONS 'NOBODY (CONS 'IS '(PERFECT))) → ?
   f. (CONS (CAR '((CAR A) (CDR A))) (CAR '(((CONS A B))))) → ?

3 ANOTHER POINT OF VIEW ON LISTS

If you still have problems with list notation, imagine them as *trees*: you go down one level each time you meet an opening parenthesis, and you climb back up a level each time you meet a closing parenthesis. Here are a few examples of lists with their tree equivalents:

The list (A B) may be represented as:

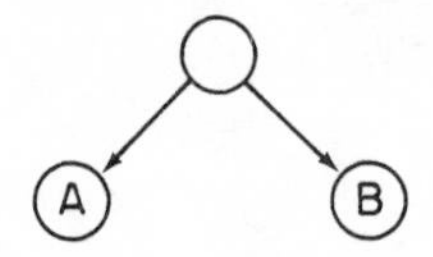

The list ((1 2 3)B) may be represented as:

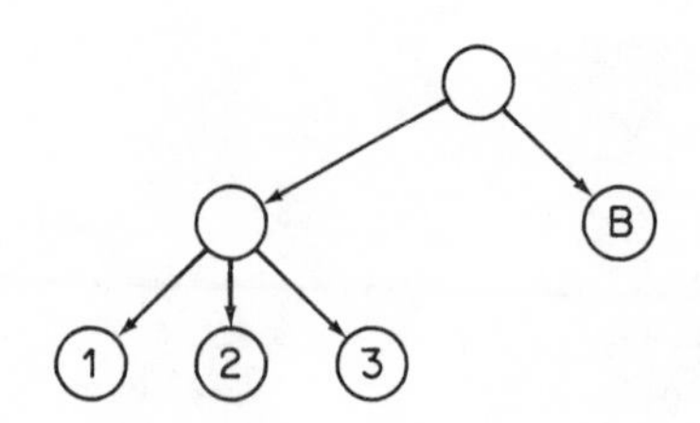

This representation allows both elements and their depths to be recognized visually. A list (or sub-list) is represented as an empty circle. The elements of a list are circles directly connected to the list circle, and the depth of an element can be easily measured by the number of arrows you have to follow (starting from the summit) to arrive at the element.

The list ((A B) (C D) (E F)) can be represented as:

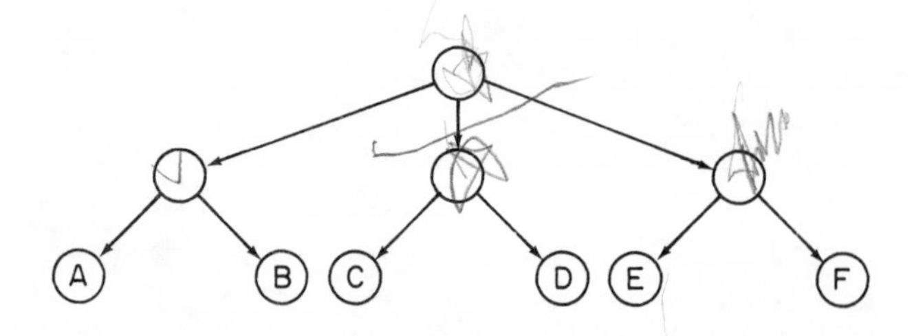

The list (((A) (B) (C)) G (((((D E)) F) H))) may be represented as:

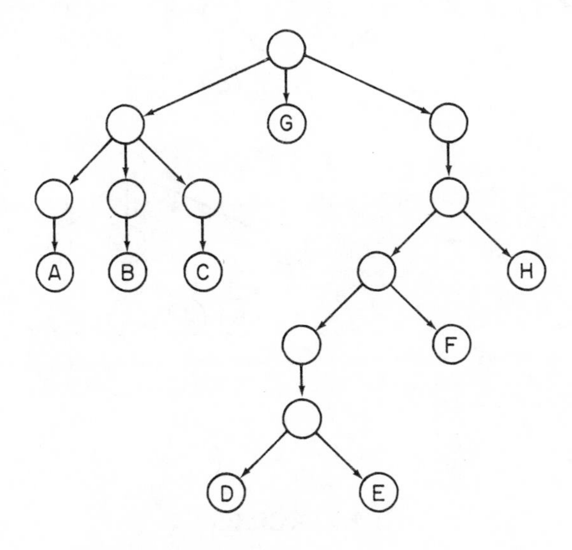

If lists are represented as trees, it is easy to visualize the two parts of a list, CAR and CDR. The CAR of a list is all the left hand branch of the tree-list, and the CDR of a list is the tree *without* the left hand branch. Here is a list:

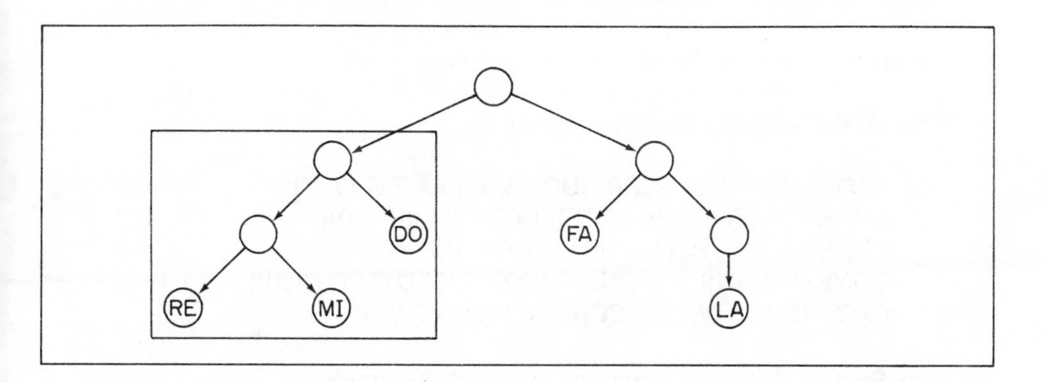

This tree is equivalent to the list (((RE MI) DO) (FA (LA))). The CAR of this list corresponds to the left hand branch contained in the internal box and the CDR corresponds to the tree that remains if the internal box is removed from the external box, i.e. if the left hand branch is removed from the complete tree. It is therefore clear that the operation CONS on tree-lists corresponds to the insertion of a new left hand branch into a tree, as below:

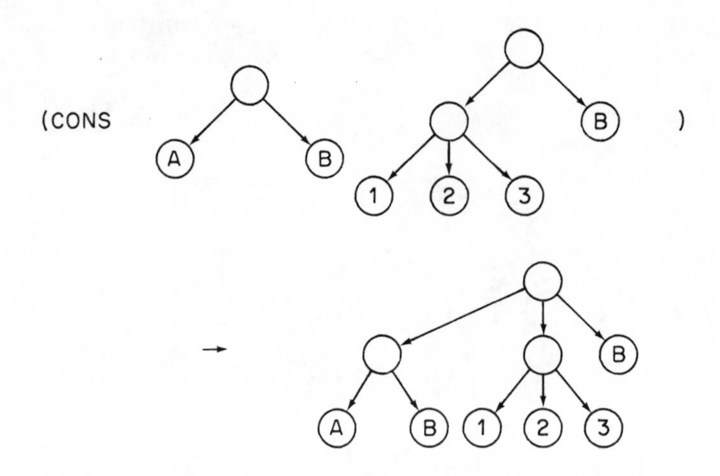

3.1 EXERCISES

1. Give the combinations of CAR and CDR necessary to replace the '?' sign in the following function calls

```
(? '(A B C D))                    → D
(? '((A (B C))E))                 → C
(? '(((GOD) STILL) ONE))          → GOD
(? '(((GOD) STILL) ONE))          → STILL
```

2. Represent the lists in the first exercise as trees.

3. What do the following function calls do:

```
(CADR (CDR (CDR (CDR '(DO RE MI FA SO LA TI)))))      → ?
(CONS (CADR '((A B) (C D)))(CDDR '(A (B(C)))))        → ?
(CONS (CONS 'HELLO NIL) '(HOW ARE YOU))               → ?
(CONS 'I (CONS 'I (CONS 'I (CONS 'STUTTER NIL))))     → ?
(CADR (CONS 'WELL (CONS '(THIS IS EASY)())))          → ?
```

4. Show the tree that results from the following operations:

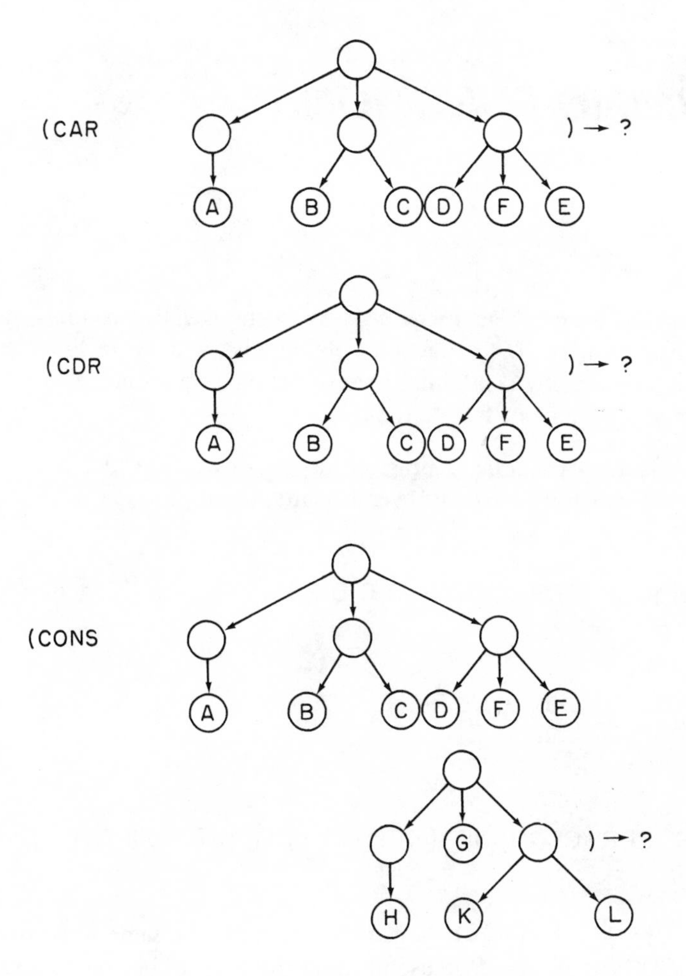

5. Represent the trees in the preceding exercise in the form of lists.

4 DEFINING FUNCTIONS

We have seen that we can construct a list (using the CONS function), take the first element of a list (using the CAR function), obtain a list without its first element (using the CDR function) and, by combining these functions, access any element in a list or combine elements at random from any lists.

Let us now consider a program (a program, as far as we are concerned, is a series of function calls) in which we have, at several points, to find the fourth element of a list, as for example in

```
(CAR (CDR (CDR (CDR '(A B C D E)))))
```

or

```
(CAR (CDR (CDR (CDR '(1 2 3 4 5)))))
```

or

```
(CAR (CDR (CDR (CDR '((0) (1) (1 0) (1 1) (1 0 0) (1 0 1) (1 1 0) (1 1
1))))))
```

Clearly, after a while it becomes tedious to keep typing the same series of calls, (CAR (CDR (CDR (CDR. . . It would be useful, in such cases, to be able to abbreviate that sequence of calls: e.g. we might like to write

```
(4-TH '(A B C D E))
```

instead of

```
(CAR (CDR (CDR (CDR '(A B C D E)))))
```

LISP makes it possible to solve this problem by *defining* new functions which, once defined, are indistinguishable from existing functions. We shall call the existing functions *standard functions* and those which the LISP user defines *user functions*.

Let us start by looking at how new functions may be defined. Here is an example of the definition of a function:

```
(DEFUN 4-TH (L)
        (CAR (CDR (CDR (CDR (CDR L)))))
```

DEFUN is a standard function indicating to the machine that we are DEFining a new function. Naturally, to be able to use this function we must give it a *name*. Any name may be used, but it is best to choose a mnemonic one, i.e. a name which reminds the user of what the function is supposed to do. In the example given, we chose the name 4-TH, which says clearly what the function is intended to do: find the fourth element of a list. Since we are dealing with a list, we have to give the function an *argument*. There is only one argument here, as we only want to calculate the fourth element of *one* list at a time. In this example, L is the name of the argument. It is a parameter of the function, i.e. the name of a *variable*. The second line of the definition of the user function 4-TH indicates what the function does: it calculates the CAR of the CDR of the CDR of the CDR of its argument L.

After this informal example, let us now look at the syntactic definition of the standard function DEFUN[1]

(DEFUN *name-of-function* (var_1 var_2 . . . var_n) *body-of-function*)

where

name-of-function
is a new name, not yet present in the machine. It is best not to take names such as CAR, CDR or CONS or more generally, names of standard functions, since if you define a function whose name is that of one of the standard functions, you will lose the original definitions.

var_1, var_2 etc
These are parameter names. Here, too, it is best to choose mnemonic names.

body-of-function
This consists of the instructions describing the operations that the function is to carry out.

The first argument of the function DEFUN is therefore the name of the function to define, the second argument is a list of parameters (did you notice the use of parentheses in the syntactic definition of DEFUN?). The rest of the definition of the function is its body, a series of calls to standard LISP functions or user functions (which must also be defined before a call to the function is executed). The value returned by a function definition (by calling the function DEFUN) is the *name* of the function defined. The definition of function 4-TH therefore returns the atom 4-TH as value. This is a message from the LISP language, showing that it has recorded your definition and that you may use it from now on.

Once the function has been *defined*, it may be *called* (to test it or execute it). A call to function 4-TH would be as follows:

```
(4-TH '(A B C D E F))
```

[1] In Le_LISP the name of the function defining function is DE. Its definition is identical to the Common-LISP function DEFUN.

The value returned by a call to a user function is the value returned by evaluating the body of the function *with the parameter variables bound to the values of the arguments of the call*. The argument for the call in the example above was (QUOTE (A B C D E F)). Parameter L in function 4-TH is therefore bound to the list (A B C D E F), which is in turn the evaluation value of the call to function QUOTE with argument (A B C D E F).

This may seem a little cumbersome (written down on paper!). It is nevertheless important to distinguish the *definition* of a function from its *call*.

The result of the call

```
(4-TH '(A B C D E F))
```

will therefore be the value of the expression

```
(CAR (CDR (CDR (CDR L))))
```

with L bound to the list (A B C D E F). This is equivalent to evaluating the expression

```
(CAR (CDR (CDR (CDR '(A B C D E F)))))
```

and therefore:

```
(4-TH '(A B C D E F)) → D
```

Here are two further calls and the results of those calls:

```
(4-TH '(1 2 3 4 5)) → 4
(4-TH '((0) (1) (1 0) (1 1) (1 0 0) (1 0 1) (1 1 0) (1 1 1))) →   (1 1)
```

Here now is a function which constructs a list from its three arguments:

```
(DEFUN LIST3 (ARG1 ARG2 ARG3)
                        (CONS ARG1 (CONS ARG2 (CONS ARG 3 () ))))
```

and here is a call to this function:

```
(LIST3 1 2 3) → (1 2 3)
```

Remember that this function is evaluated in the following order:

- variable ARG1 is first bound to value 1
- next variable ARG2 is bound to value 2
- next variable ARG3 is bound to value 3

● finally the body of the function is executed.

Each time there is a reference to one of the three variables ARG1, ARG2 or ARG3, the LISP language calculates their respective value. The value of a variable is the value bound to it.

And here are a further two calls to this function:

```
(LIST3 'WELL 'WELL 'WELL)                → (WELL WELL WELL)
(LIST3 'YET '(ANOTHER LIST) 'OK?)        → (YET (ANOTHER LIST) OK?)
```

What does the following function do:

```
(DEFUN BUILD (SOMETHING)
      (CONS SOMETHING (CONS SOMETHING (CONS SOMETHING ()))))
```

here is a call to the function:

```
(BUILD 'HMMM) → ?
```

Let us examine this step by step: this is a function called BUILD, which on being called takes an argument which will be bound to the variable called SOMETHING. Consequently, when function BUILD is called, LISP binds variable SOMETHING to the argument HMMM, which is the value of expression (QUOTE HMMM). Next, every occurrence of variable SOMETHING within the body of the function is replaced by this value. This gives the same as if we had written:

```
(CONS 'HMMM (CONS 'HMMM (CONS 'HMMM())))
```

Next, we have only to evaluate this expression. As always, LISP first evaluates the argument of standard functions and only then is able to calculate the value of the function itself. Let us follow this in detail:

1. The first thing to do is to evaluate the two arguments of the first CONS: the expression 'HMMM, the first argument, and (CONS 'HMMM (CONS 'HMMM ())))), which is the second argument. The evaluation of 'HMMM simply gives the atom HMMM. This atom will be the new first element of the list resulting from the evaluation of (CONS 'HMMM (CONS 'HMMM ())).

2. To calculate this list, we have to carry out a CONS between 'HMMM and (CONS 'HMMM ()). As before, the value of the expression (QUOTE HMMM) is the atom HMMM. This atom wil be the new first element of the list resulting from the evaluation of (CONS 'HMMM()).

3. The next stage is simply to put a pair of parentheses around the atom HMMM giving the list (HMMM).

4. We can now calculate the CONS that we have left in abeyance at stage 2. We have to calculate the result of a CONS of HMMM and (HMMM). This gives the list (HMMM HMMM).

5. There was still a CONS to calculate in stage 1: the CONS of atom HMMM and list (HMMM HMMM). The final result is therefore the list (HMMM HMMM HMMM)

The function BUILD therefore returns a list with three occurrences of the element passed as argument. This is very powerful: having defined a function it is possible to use it in the same way as a standard function.

If you have any problems with understanding this section, re-read it carefully, as it is *very* important for the rest of our discussion, and do the exercises below.

4.1 EXERCISES

1. Here is the definition of a very interesting function:

```
(DEFUN QUICK (L1 L2)
  (CONS
    (CONS (CAR L1) (CDR L2))
    (CONS (CAR L2) (CDR L1))))
```

   What will it do for the following calls:

```
(QUICK '(A 2 3) '(1 B C))                    → ?
(QUICK '(JA CA VA) '(OUI ES GEHT))           → ?
(QUICK '(A VOITURE) '(UNE CAR))              → ?
```

2. What are the results of the following function calls?

```
(BUILD (4-TH '(DO RE DO MI DO FA)))          → ?
(QUICK (LIST3 'UN 'CUBE 'ROUGE)
(QUICK (LIST3 'SUR 'LA 'TABLE) (BUILD 'BRAVO))   → ?
```

3. Write a zero argument function which returns the following list when called:

```
(HELLO)
```

4. Write a one-argument function which returns a list containing four occurrences of this argument.

5. Write a three-argument function which returns a list containing the three arguments with their order inverted. For example: if the function is called REVERSE3 a possible call might be:

```
(REVERSE3 'DO 'RE 'MI)
```

and the value returned would be (MI RE DO).

5 PREDICATES AND SELECTION

With the functions now available to us, it is already possible to write a large number of useful small programs. Such programs, however, are *linear*. By this we mean that they are made up of a series of instructions (a series of function calls) which will be executed one after another, without tests or repetitions. This chapter introduces particular functions, called predicates, and a *selection* or *test* function which uses these predicates.

Remember that programs may be regarded as descriptions of activities, rather like recipes. As with recipes, we have to be able to test the state of the world, e.g. it must be possible to find out whether the water is boiling or not. Test functions in logic are called *predicates*. These are functions asking questions to which the reply will be either yes or no. They test whether a condition is met, and therefore may return one of *two* possible values: *yes* or *no*, as in the question 'is the water boiling?' to which you could only reply yes or no (the rules of the game forbid answers of the 'not yet' type). The machine may only reply yes or no, *true* or *false*, when evaluating a predicate.

In LISP, *no* is expressed by NIL or (), and *yes* by T. T is an abbreviation of True, and nil in both Latin and colloquial English, means nothing, false, null. This association between the words *false* and the word *nothing* appears in LISP too: we should remember that NIL is also used for the empty list (). NIL therefore plays several roles in LISP:

- it is an atom. NIL is an atom, but not a variable: no value can be bound to it.

- it is a name for the empty list (a list with 0 elements)

- it is the LISP word for *logically false*.

5.1 A FEW PREDICATES

Here is the syntactic definition of a first predicate – the function ATOM:

(ATOM *arg*) → () if *arg* is not an atom
→ T if the argument is an atom

This function therefore tests to see whether the value of an argument is an atom. Here are a few examples:

```
(ATOM 'BRAVO)                          → T
(ATOM NIL)                             → T
```

Note that there is no need to *quote* the atom NIL. In the same way, there is no need to quote atom T and the numbers. Such atoms are called *constants*.

```
(ATOM ())                              → T ;by definition;
(ATOM '(A B))                          → () ;this is a list;
(ATOM 1024)                            → T
(ATOM '(GIRLS (ARE (COSIER))))         → ()
```

We have a predicate which tests whether an argument is an atom, and there are obviously predicates which test other types of LISP objects. For example, the predicate NUMBERP tests whether its argument is a *number* and the predicate CONSP tests whether its argument is a *List*. Here are their syntactic definitions and a few examples of calls:

```
(CONSP arg)          → NIL if arg is not a list
                     → T if the argument is a list
```

a few examples of (CONSP:

```
(CONSP '(A B C))               → T
(CONSP 'BRAVO)                 → NIL
(CONSP ())                     → NIL; by definition;
(CONSP '((A) (B C)))           → T
(CONSP -1024)                  → NIL
```

and here is the predicate NUMBERP:

```
(NUMBERP arg)        → NIL if arg is not a number
                     → T or arg if the argument is a number
```

A comment needs to be made at this stage: we said a little while ago that in LISP *true* is translated by T. This is not entirely correct, as we should have said that *false* is translated by NIL in LISP, but LISP considers any expression differing from NIL as equivalent to T, i.e. as *true*. This is why in function NUMBERP the result may be either NIL if the argument is not a number, or the value of the argument itself if it is a number. In Le_LISP, all predicates that allow the argument of the call itself to be returned as a *true* value will stick to this convention, which is very useful if the value of the argument is to be used later in the program. In Common-LISP, the convention is that *all* predicates return either T or NIL. In this book, we will never make use of Le_LISP's possibility to return either NIL or – possibly – parts of the argument. Nevertheless, in the following examples we will – from time to time – show the Le_LISP result, so that you get used to the fact that everything different from

NIL is considered as logically true. We shall return to this question when the occasion arises. In any case, the convention is not possible for function ATOM, since the call

(ATOM NIL)

must return a *true* value, i.e. T. It cannot reasonably, in this case, return the value of the argument, which is NIL, the LISP value saying 'logically false'.

Here are a few examples of calls to function NUMBERP:

```
(NUMBERP 0)           → 0
(NUMBERP −110)        → T
(NUMBERP NIL)         → ()
(NUMBERP 'HMMM)       → ()
(NUMBERP 3214)        → T
(NUMBERP '(DO RE))    → ()
```

Let us at once introduce a further predicate: function NULL. This function tests whether its argument is an empty list or not. Here is the definition:

(NULL *arg*) → NIL if *arg* is not an empty list
→ T is the argument is the empty list.

NULL can also be viewed as expressing the question "is it true that the value of the argument is equal to NIL or () ?". Here are a few examples:

```
(NULL ())             → T
(NULL NIL)            → T
(NULL T)              → ()
(NULL'(WELL))         → ()
(NULL 13)             → ()
(NULL 'AN-ATOM)       → ()
(NULL (NULL NIL))     → ()
(NULL (NULL T))       → T
```

Clearly since NIL is the LISP equivalent of logical *false*, this predicate may also be treated as saying 'Is it true that the value of the argument is false?' It is an extremely useful function, as we shall see in the next chapter.

These predicates may be used in functions like any other LISP functions. Let us for example write a function which tests whether the CAR (speaking loosely: CAR means the first element) of a list is an atom:

```
(DEFUN ATOM-CAR? (L)
       (ATOM (CAR L)))
```

If we give this definition of a function to LISP, it will reply by returning the atom ATOM-CAR?, indicating that it now knows this function. We can then call the function:

```
(ATOM-CAR? '((DO RE) (RE DO)))              → NIL
or
(ATOM-CAR? '(DO RE RE DO))                  → T
or again
(ATOM-CAR? (CONS 'HELLO '(YOU THERE)))      → T
```

5.1.1 Exercises

1. Write a function NUMBERP-CADR, which tests whether the second element of the list passed as argument is a number.

2. Write a function called LIST-CAR? which returns the list (IS x), with x equal either to T or to NIL, depending on whether the first element of the list of the functions arguments is a list or not. Here are some examples of calls:

```
(LIST-CAR? '(A B C))          → (IS NIL)
(LIST-CAR? '((A) (B C)))      → (IS T)
(LIST-CAR? '(AHA))            → (IS NIL)
```

3. What are the results of these function calls:

```
(NUMBERP-CADR '(123))                   → ?
(NUMBERP-CADR '(A-24 B))                → ?
(NUMBERP-CADR '(256 LITTLE BITS)        → ?
(NUMBERP-CADR (CONS '1 '(12358 13)))    → ?
```

with NUMBERP-CADR defined as follows:

```
(DEFUN NUMBERP-CADR (L)
    (NUMBERP (CADR (LIST3 (CADDDR L) (CAR L) (CADR L)))))
```

and LIST3 defined as in the previous chapter.

5.2 THE IF FUNCTION

Let us return to our analogy with recipes (or an instruction manual): if we test the temperature of the water, i.e. if we ask the question 'Is the water boiling?' it is probably because we intend to *act differently* depending on the result of the test. We might, for example, put out the heat or pour the water into a teapot if it is boiling, and wait for it to boil otherwise. Consequently, we need to choose the appropriate activity from several, depending on the results of the application of the test.

In LISP, predicates work in exactly the same way: they guide the subsequent execution of a program depending on the results of tests on data.

The LISP function responsible for this 'guiding' is the IF function. Here is its syntactic definition:[1]

(IF *test* *action-if-true*
        $action\text{-}if\text{-}false_1$
        $action\text{-}if\text{-}false_2$
        . . .
        $action\text{-}if\text{-}false_n$)
  → value of *action-if-true* if the test evaluates to *true*
  → value of $action\text{-}if\text{-}false_n$ if the test evaluates to *false*

This definition can be interpreted in the following way:

> IF is a function with a variable number of arguments. The first argument, *test*, is a predicate. The second argument, *action-if-true*, is any LISP expression. If the result of *test* is different from NIL (i.e. if the test gives the result *true*) the value returned by IF will be the value of the expression *action-if-true*. If, on the other hand, the result of evaluating the *test* is equal to NIL, the LISP machine evaluates one *action-if-false* after the other and returns as value the result of evaluating $action\text{-}if\text{-}false_n$.

Let us look at an example. Here is a possible call to IF:

```
(IF (NUMBERP 1) '(A NUMBER) '(NOT A NUMBER)) → (A NUMBER)
```

Evaluation of the test (NUMBERP 1) returns as value T, which is clearly different from NIL. Consequently, the result of evaluating the *test* gave *true*. This implies, according to the definition of IF, that the value returned by the IF as a whole will be the value of *action-if-true*, and therefore the value of '(A NUMBER), i.e. the list (A NUMBER). We could have had a different result if we had submitted the following call to LISP instead of (NUMBERP 1):

```
(IF (NUMBERP 'HELLO) '(A NUMBER) '(NOT A NUMBER)) → (NOT A NUMBER)
```

Since atom HELLO is not a number, the value returned is the list (NOT A NUMBER), the value of the only *action-if-false* element present.

Here are a few more examples:

[1] Some implementations of LISP, such as Common-LISP, accept only *one* action when the evaluation of *test* yields NIL.

```
(IF (NULL '(A B)) (CAR '(DO RE))(CDR '(DO RE)))          → (RE)
(IF (CONSP (CONS NIL NIL)) (CONS NIL NIL) 'BRRR)         → (NIL)
(IF NIL 1 2)                                             → 2
```

in the latter example, the first argument of the IF was not a predicate call. Nevertheless, the atom NIL plays the role of a predicate, simply by virtue of the fact that it is in the *test position*. The result of the evaluation of *test* is therefore equal to ().

Here is yet another example:

```
(IF '(A B C) 'AHA 'HMMM) → AHA
```

the same comment applies as before: the list (A B C) is not a predicate, but because it is in the *test* position, it plays the same role as a logical *true*.

With this function available to us, we can finally begin to write real (small) LISP functions. Let us start by building a new function NUMBERP-CADR? which returns the list (second-element IS A NUMBER) if the second element of its argument list is a number, and otherwise returns the list (second-element IS NOT A NUMBER):

```
(DEFUN NUMBERP-CADR? (L)
    (IF (NUMBERP (CADR L))
        (CONS (CADR L) '(IS A NUMBER))
        (CONS (CADR L) '(IS NOT A NUMBER))))
```

and here are a few calls to the function:

```
(NUMBERP-CADR? '(1 2 3))          → (2 IS A NUMBER)
(NUMBERP-CADR? '(DO RE MI))       → (RE IS NOT A NUMBER)
(NUMBERP-CADR? NIL)               → (NIL IS NOT A NUMBER)
```

and a few more examples:

```
(DEFUN CONS? (ELE L)
(IF (CONSP L) (CONS ELE L) (CONS L '(IS NOT A LIST, BUDDY))))
```

and here are a few calls:

```
(CONS? 1 '(2 3 4))
        → (1 2 3 4)
(CONS? 1 2)
        → (2 IS NOT A LIST, BUDDY)
(CONS? 'A (CONS? 'GREEN (CONS? 'MOUSE ())))
        → (A GREEN MOUSE)
(CONS? 'A (CONS? 'MOUSE 'GREEN))
        → (A GREEN IS NOT A LIST, BUDDY)
```

Now consider the following function:

```
(DEFUN TYPE? (ARG)
        (IF (NUMBERP ARG) 'NUMBER
            (IF (ATOM ARG) 'ATOM 'LIST)))
```

As every argument in an IF can be any LISP expression, nothing prevents one or more of these arguments being calls to another IF function. Nesting of this kind is generally used to distinguish between more than two possibilities. Here, in the function TYPE?, we distinguish three possible cases: either the argument given in the call is a number, and the machine replies with the value of the NUMBER, or it is an atom, in which case the machine returns the atom ATOM, or the argument is neither a number nor an atom, in which case the machine assumes that it is a list and returns the atom LIST as value. Here are a few calls:

```
(TYPE? '(((IRREN) IST) MENSCHLICH))          → LIST
(TYPE? '(ERRARE HUMANUM EST))                → LIST
(TYPE? (CAR '(1 + 2)))                       → NUMBER
(TYPE? (CADR '(1 + 2)))                      → ATOM
(TYPE? '(THIS IS FALSE))                     → LIST
```

5.2.1 Exercises

1. Write a function 3NUMBERS which returns the atom BRAVO, if the first three elements of its argument list are numbers, and otherwise returns the atom LOSE; e.g.:

```
(3NUMBERS '(1 2 3))                          → BRAVO
(3NUMBERS '(1 CAT))                          → LOSE
(3NUMBERS (CONS 1 (CONS -100 (CONS -1 ()))))→ BRAVO
```

2. Write the function REV which inverts a list of 1, 2 or 3 elements. Examples of calls are:

```
(REV '(DO RE MI))                            → (MI DO RE)
(REV '(IS HE TIRED))                         → (TIRED HE IS)
(REV '(JO GEHNS))                            → (GEHNS JO)
```

3. Here is a weird function:

```
(DEFUN WEIRD (ARG1 ARG2)
        (IF (CONSP ARG2)
            (CONS ARG1 ARG2)
            (IF (CONSP ARG1)
                (CONSP ARG2 ARG1)
                (IF (NULL ARG2)
                  (IF (NULL ARG1) '(WELL THEN)
                           (CONS ARG1 ARG2))
                  (IF (NULL ARG1) (CONS ARG2 ARG1)
                    (CONS ARG1 (CONS ARG2 NIL)))))))
```

What does this function do? Give the values of the following calls:

```
(WEIRD () ())                                                     → ?
(WEIRD 1 '(2 3))                                                  → ?
(WEIRD '(2 3) 1)                                                  → ?
(WEIRD 'SO 'ITS YOU)                                              → ?
(WEIRD () (TYPE? 432))                                            → ?
(WEIRD (CDR (CONS? (TYPE? 'INTEL) 432)) (CAR '(DO RE MI)))        → ?
```

Before going on, make sure you have understood everything so far. If you have problems, go over the exercises and the examples again and do not forget to run them on a machine, check them, modify them slightly and hunt down any errors.

6 REPETITION

6.1 THE REVERSE FUNCTION

Any process description language, and therefore any programming language, must include the possibility of describing repetitive activity. LISP expresses repetition in a particularly elegant, easy and efficient way. Let us look first at an example of a function which inverts the elements of any list:

```
(DEFUN REVERSE (L RES)
       (IF (NULL L) RES
          (REVERSE (CDR L) (CONS (CAR L) RES))))
```

and here are a few calls:

```
(REVERSE '(DO RE MI FA) NIL)                    → (FA MI RE DO)
(REVERSE '(COLOURLESS GREEN IDEAS) NIL)    →
                                      (IDEAS GREEN COLOURLESS)
(REVERSE '(A B (C D) E F)NIL)                   → (F E (C D) B A)
```

To make clear how a call to this function is evaluated, let us first give a description of its behaviour: imagine that we have two areas, one called L, the other called RES, each in an initial state:

L	RES
(A B C D)	()

To invert the list L consisting of (A B C D), we shall transfer one element after the other from L towards RES, taking each element as it leaves L and adding it to the head of RES. Here is a graphic *trace* of the successive states of L and RES:

	L	RES
Initial state	(A B C D)	( )
Operation		
First transaction	(B C D)	(A)
Operation		
Second transaction	(C D)	(B A)
Operation		
Third transaction	(D)	(· C B A)
Operation		
Final transaction	()	(D C B A)

The REVERSE program behaves in exactly the same way: it transfers each element of L in turn to the accumulator RES, and halts when list L is emptied.

It is easy to understand the writing of this function if we think about it as follows: to invert a list, we first have to check whether the list is empty. If yes, then there is nothing further to be done and the result is the list that has been accumulated in variable RES. Moreover, this is naturally true at the start, as RES will be equal to NIL, the empty list, when first called. The empty list is indeed the reverse of the empty list. The next stage is to take the first element of list L and to put it at the head of list RES.

We then have to go on doing the same thing with the remaining elements, *until list* L *is empty*. If we could now find a function to do this task, we would merely have to call it. But we do have a function that meets this requirement: function REVERSE which we are currently defining. Consequently, let us call that function for the rest of list L and the accumulated contents, in RES, of the first element of L and of RES. Note that it is necessary to modify the argument L in such a way that it tends towards satisfying the halt test (NULL L).

Executing the call

(REVERSE '(DO RE MI) NIL)

therefore takes place as follows:

1. First, L is bound to the list (DO RE MI) and RES to NIL. Next the body of the function is evaluated with these bindings. The first thing to do is to test whether L is empty. For the moment it is not. We must therefore evaluate the *action-if-false* part of the IF function, which tells us that (REVERSE (CDR L) (CONS (CAR L) RES)) must be calculated. With the bindings still valid for the moment, this means that we have to calculate

   (REVERSE '(RE MI) '(DO))

To find the result of the initial call, we therefore only need to know the result of this new call.

2. We now apply function REVERSE again, binding variable L to the new value (RE MI) and the variable RES to (DO). As L is still not equal to NIL, we will once more have to evaluate *action-if-false* in the selection to find the result of this call. This time it will be

   (REVERSE '(MI) '(RE DO))

   i.e. the CDR of L and the result of (CONS (CAR L) RES) with the current associations.

3. The same operation takes place again: new bindings are made for the variables, and L is still not equal to NIL we call REVERSE again.

   (REVERSE () '(MI RE DO))

4. This time variable L is bound to NIL and the test (NULL L) is therefore true. The result of the final call to REVERSE is therefore the value of RES, the list (MI RE DO). But remember that this call only took place in order to know the result of the preceding call, and it is therefore also the result of that preceding call. Consequently, working backwards to the initial call, the machine can give us that list as the value returned by the function as a whole.

We can represent the evaluation of this call graphically as follows:

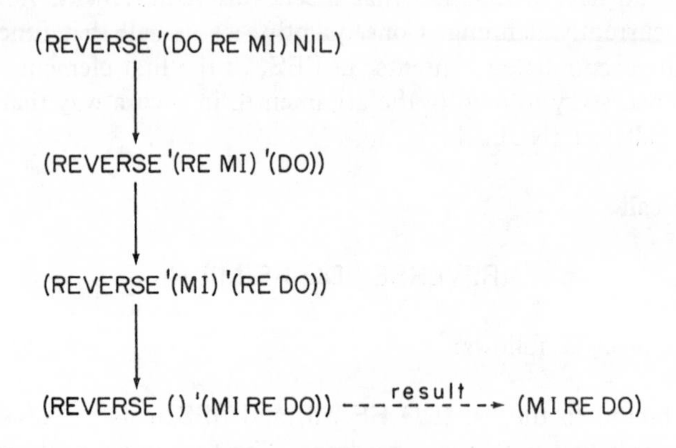

Functions containing a call to themselves within their *body* are called *recursive functions*. Function reverse is therefore a recursive function.

Here is a *trace* of the execution of the same call:

```
?(REVERSE '(DO RE MI) ())
      → REVERSE: ((DO RE MI) NIL)
        → REVERSE: ((RE MI) (DO))
          → REVERSE: ((MI) (RE DO))
            → REVERSE: (() (MI RE DO))
←────────── REVERSE = (MI RE DO)
= (MI RE DO)
```

6.2 THE APPEND FUNCTION

Let us consider a second example of a recursive function: function APPEND which concatenates two lists as in the following examples:

```
(APPEND '(A B) '(C D))            → (A B C D)
(APPEND '(A) '(B C))              → (A B C)
(APPEND NIL '(A B C))             → (A B C)
```

This function has two parameters (for the two lists passed as arguments) and returns a list which is the concatenation of the other two. This provides us with an initial definition of the function:

```
(DEFUN APPEND (LIST1 LIST2)
                              . . .
```

If we look at the examples of the APPEND function (these examples may be taken as its *specification*), we see that if the first argument, LIST1, is empty, the result is the list passed as second argument, i.e. LIST2, and if LIST1 has only one element, the result is LIST2 with a new first element, the single element in the first argument. Unfortunately, we already know that we cannot apply CONS to the second list and each element in turn from the first, as this method was what we used for function REVERSE: it would add the elements in inverse order. We need therefore to be able to begin with the final element of list LIST1, add it to list LIST2, and then take the second to last element of LIST1 and so on, until we reach the first element of LIST1. To do this, we could start by inverting LIST1 and then apply the REVERSE function with the inverted list and LIST2 as arguments. This would therefore give us:

```
(DEFUN APPEND (LIST1 LIST2)
        (REVERSE (REVERSE LIST1 NIL) LIST2))
```

But, though this way of writing the function is very brief, it is extremely inefficient since LIST1 is completely scanned twice. Think of the time that this would take if we were using a very long list.

Exactly the same thing can be done more efficiently, if we separate the scanning of LIST1 from the construction of the result list: start by scanning the first argument list, LIST1, storing the different elements met, and then carry out a CONS of each element, starting with the last, and the second argument list, LIST2. This gives us a second definition:

```
(DEFUN APPEND (LIST1 LIST2)
       (IF (NULL LIST1) LIST2
              (CONS (CAR LIST1) (APPEND (CDR LIST1) LIST2))))
```

which may be understood as follows:

To concatenate two lists, two special cases have to be distinguished:

1. If the first list is empty, the result is the second list.

2. Otherwise, we only need to add the first element of the first list to the head of the list produced by concatenating the rest of the first list with the second list. This is a natural enough operation, since we have already seen that concatenation of a list with one element is quite simply the result of a CONS of this single element and the second list.

Graphically, execution of the call

```
(APPEND '(A B) '(D E))
```

can be represented as follows:

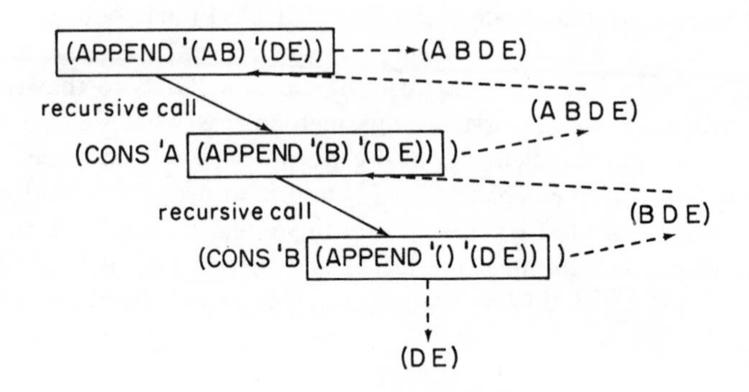

Clearly, as we move *downwards* through the successive recursive calls, the machine accumulates CONS functions which it can only work out after having evaluated their arguments (here, one of the arguments is always a recursive call to APPEND). Calculating the CONS expressions is carried out as we climb *back up* from the recursions.

Here is a *trace* of a call to function APPEND:

```
?(APPEND '(A B C) '(D E F))                     ;initial call;
   → APPEND: ((A B C) (D E F))                    ;its trace;
    → APPEND: ((B C) (D E F))                   ;second call;
                                  ;i.e. the first recursive call;
       → APPEND:((C) (D E F))                     ;third call;
          → APPEND:(NIL (D E F))                 ;fourth call;
         ← APPEND = (D E F)             ;result of fourth call;
       ← APPEND = (C D E F)              ;result of third call;
      ← APPEND = (B C D E F)            ;result of second call;
    ← APPEND = (A B C D E F)             ;result of first call;
= (A B C D E F)                        ;result of initial call;
```

Comparing the *traces* for REVERSE and APPEND, we find that there are two types of recursive loops:

1. Recursive loops which carry out all the computations as they go *downwards* i.e.: as in function REVERSE, where computation takes place entirely through recursive calls. Once the *exit test* from recursion has been reached (here, the test is the expression (NULL L)), computation is complete and the function can simply return the value directly.

2. Recursive loops which carry out part of the computation on the way *down* the recursive calls and others while climbing back *up*. This is the case of function APPEND: as it moves downwards, the function identifies the different elements which it will combine with the results of the recursive calls later, by applying the CONS function to them, on the way back up. Note that the elements separated out during the downward movement are used, while climbing back upwards, in reverse order, i.e. the first element separated out will be the last element used.

We shall return later to the difference between these two types of recursive functions. For the moment, let us simply say that these two types are easy to distinguish in the way functions are actually written: in the first type (computation taking place only on the way down) the result of a recursive call is not used by another function, or, more precisely, the recursive call is not the argument of a function; in the second type (computation both on the way down and the way up), results are used by other functions. In function APPEND, for example, the result of each recursive call is used as the second argument of a CONS.

Calls which lead to computation only on the way down are called *terminal recursive calls*.

6.3 FUNCTION EQUAL

Our third example of a recursive function is the function EQUAL, which tests the equality of its two arguments. We have to start by learning a new predicate: predicate EQ which tests the equality of two *atoms*. Here is a definition of EQ:

(EQ *arg1 arg2*) →T if *arg1* and *arg2* are the same atom

→NIL if *arg1* and *arg2* are different atoms or if at least one of them is a list.

Here are a few examples of the use of the standard function EQ:

```
(EQ 'LUKE 'LUKE)                                  → T
(EQ 'LUKE 'DANIEL)                                → NIL
(EQ (CAR '(DO RE)) (CADR '(MI DO SOL)))           → T
(EQ '(D P F T R) '(D P F T R))                    → NIL
```

Note that EQ returns NIL, and therefore *false*, if the arguments are lists, even if the two lists are equal[1]. There is no predefined function to test the equality of two lists. Moreover, we know that two list may be regarded as equal if and only if corresponding pairs of elements from the two lists are equal. All we therefore have to do is to compare the successive elements of two lists by applying function EQ. This leads to the following definition:

```
(DEFUN EQUAL (ARG1 ARG2)
       (IF (ATOM ARG1) (EQ ARG1 ARG2)
          (IF (ATOM ARG2) NIL
             (IF (EQ (CAR ARG1) (CAR ARG2))
                (EQUAL (CDR ARG1) (CDR ARG2))
                NIL))))
```

Clearly, this function tests both the equality of two atoms (the first line says that if ARG1 is an atom, all we need to do to test its equality – by a call to EQ – to ARG2) and the equality of two lists (testing first of all the equality of the first two elements and then repeating the process with the CDRs of the two lists).

Note the use of predicate ATOM: it is used both to test if we are dealing with an atom *and* provides the halt test for recursivity. This is possible since the empty list is regarded as the atom NIL. Here are a few examples of calls to this function:

```
(EQUAL '(DO RE MI) '(DO RE MI))                                  → T
(EQUAL'WELL'WELL)                                                → T
(EQUAL '(DO RE MI) (CDR '(DO RE MI)))                            → NIL
(EQUAL '(LE NEZ (DE CLEOPATRE)) '(LE NEZ (DE CLEOPATRE)))→ NIL[2]
```

Clearly, the function does not work if the lists given as arguments contain sub-lists. Here is a trace of the last example:

[1] The exception to this rule will be discussed in Chapter 15.

[2] To find out more about 'le nez de Cleopatre' (Cleopatra's nose) readers should consult Raymond Queneau's wonderful article on *litterature definitionnelle* in *Oulipo, la litterature potentielle*, pp. 119–122, in the Gallimard *idées* series.

```
?(EQUAL '(LE NEZ(DE CLEOPATRE)) '(LE NEZ DE CLEOPATRE)))
  → EQUAL: ((LE NEZ(DE CLEOPATRE))(LE NEZ(DE CLEOPATRE)))
    → EQUAL: ((NEZ(DE CLEOPATRE))(NEZ (DE CLEOPATRE)))
      → EQUAL: (((DE CLEOPATRE))((DE CLEOPATRE)))
  ←——— EQUAL = NIL
  = NIL
```

The test (EQ (CAR ARG 1) (CAR ARG 2)) returns NIL if the first element is a list. To overcome this failing, we need to test not equality EQ, but equality EQUAL of successive CARs. This gives us the following modified version:

```
(DEFUN EQUAL (ARG1 ARG2)
        (IF (ATOM ARG1) (EQ ARG1 ARG2)
            (IF (ATOM ARG2) NIL
                     (IF (EQUAL (CAR ARG1)(CAR ARG2))
                        (EQUAL (CDR ARG1)(CDR ARG2))
                        NIL)))
```

and here is a trace of this new function on the same example:

```
?(EQUAL '(LE NEZ(DE CLEOPATRE)) '(LE NEZ (DE CLEOPATRE)))
  → EQUAL: ((LE NEZ (DE CLEOPATRE)) (LE NEZ (DE CLEOPATRE)))
   → EQUAL: (LE LE)
   ← EQUAL = T
   → EQUAL: ((NEZ(DE CLEOPATRE))(NEZ (DE CLEOPATRE)))
     → EQUAL: (NEZ NEZ)
     ← EQUAL: T
     → EQUAL: (((DE CLEOPATRE))((DE CLEOPATRE)))
       → EQUAL: ((DE CLEOPATRE) (DE CLEOPATRE))
         → EQUAL:  (DE DE)
         ← EQUAL = T
         → EQUAL: ((CLEOPATRE)(CLEOPATRE))
           → EQUAL: (CLEOPATRE CLEOPATRE)
           ← EQUAL= T
           → EQUAL: (NIL NIL)
       ←——— EQUAL = T
       → EQUAL: (NIL NIL)
   ←——— EQUAL = T
  = T
```

Note that this function contains two recursive calls: the result of the first call is used as a test for the IF statement, and the second call gives us the result of the function. This function therefore contains a terminal recursive call (the second recursive call) and a non-terminal recursive call (the first recursive call).

A final remark on this function: it is often necessary to write a function with a set of nested IFs, as is the case here, where there are three IFs one after the other. In such cases, there is an abbreviation which is both more readable and more powerful: this is function COND, a generalized selection function. It is defined syntactically as follows:

(COND *clause*$_1$ *clause*$_2$ . . . *clause*$_n$)

and each of these *clauses* must take the form:

(*test* [*action*$_1$ *action*$_2$ . . . *action*$_n$])

The evaluation of function COND takes place as follows:

the *test* in the first clause (*clause*$_1$) is evaluated

[1] if its evaluation returns the value NIL
then if there are still other clauses
continue with the evaluation of the *test*
of the next clause and continue as in [1]
else halt evaluation of the COND
and return NIL as value
else evaluate in sequence actions *act*$_1$ to *act*$_n$ of the current clause and leave the COND returning as value the evaluation of *act*$_n$. (If there are no actions to evaluate, leave the COND returning as value the evaluation of *test*.)

[2] if none of the *tests* are satisfied, the value of COND is NIL

This means that the expression:

```
(IF test1 act1
    (IF test2 act2
        (IF test3 act3
            . . . .
              (IF testn actn1 actn2 . . .actnm). . .)))
```

is equivalent to the expression:

```
(COND
        (test1 act1)
        (test2 act2)
        (test3 act3)
                        . . .
        (testn actn1)
        (T actn2 . . . actnm))
```

The structure of COND is far more readable than a set of nested IFs. We can therefore write function EQUAL in a more elegant form (which still does exactly the same thing):

```
(DEFUN EQUAL (ARG1 ARG2)(COND
        ((ATOM ARG1) (EQ ARG1 ARG2))
        ((ATOM ARG2) NIL)
        ((EQUAL (CAR ARG1)(CAR ARG2))
                (EQUAL (CDR ARG1)(CDR ARG2)))))
```

6.4 THE DELETE FUNCTION

We shall now examine a final example of a recursive function. This time the problem is to write a function which removes all the occurrences of a given element in a list. This function, which we shall call DELETE, therefore has two arguments: the first argument indicates the element to eliminate, and the second gives the list from which the element is to be eliminated. Its construction will be *immediate*.

Remember that in all functions with repetitive structures, there is *always* at least one halt test and at least one recursive call within which at least one of the arguments must be modified in such a way that the successive values of an argument converge towards the satisfaction of the halt test.

To be able to *remove* a given element from a list, we naturally have to start by *finding* that element. The EQUAL function uses a search algorithm. Nothing prevents us using it again. The difference here is that we will not be satisfied with having found *one* occurrence of the element in question, but will go on to find all occurrences of it. Moreover, an algorithm for search alone is not sufficient, as each time we have discovered an occurrence of the element we shall want to eliminate it. In addition, we want to keep all the other elements in the list. The program can virtually be directly derived from what we have been saying: we shall compare the list element by element with a particular element; if it is different, we leave it in the list and go on looking, otherwise we remove it (what we actually do is decide not to keep it) and again go on searching for other occurrences in the rest of the list. We stop once the list is empty.

This algorithm is easily translatable into LISP:

```
(DEFUN DELETE (ELT LIST) (COND
        ((NULL LIST) NIL)
        ((EQUAL ELT (CAR LIST)) (DELETE ELT (CDR LIST)))
        (T (CONS (CAR LIST) (DELETE ELT (CDR LIST))))))
```

If element ELT does not belong to list LIST, this function returns a copy of LIST. Note too the use of function EQUAL as a predicate.

Here are a few examples of the function:

```
(DELETE 'A '(A A H))                     → (H)
(DELETE '(A) '(A A H))                   → (A A H)
(DELETE '(A) '(A (A) H))                 → (A H)
(DELETE '(DO RE) '(FA SOL (DO RE)))      → (FA SOL)
```

here is a trace of execution of a call to function DELETE. Note again the mixture of terminal and non-terminal recursive calls:

```
?(DELETE 'A'(A A H A H))
      → (DELETE A (A A H A H))
        → (DELETE A (A H A H))
          → (DELETE A (H A H))
            → (DELETE A (A H))
              → (DELETE A (H))
                → (DELETE A NIL)
                ← DELETE NIL
            ←—— DELETE (H)
   ←———————— DELETE (H H)
  = (H H)
```

6.5 EXERCISES

1. If DELETE is called as follows:

   (DELETE 'A (A (B A (C A) A) A))

   the result would be ((B A(C A)A)). Change the function so that it returns the result ((B (C))), i.e. modify the function so that it eliminates *any* occurrence of a given element, whatever the depth at which the element is located within the list.

2. Write a function which doubles each of the elements in a list given as argument. Examples of possible calls are:

```
(DOUBLE '(A B C))              → (A A B B C C)
(DOUBLE '(DO (RE MI)FA))       → (DO DO (RE MI)(RE MI) FA FA)
(DOUBLE '(I STUT TER))         → (I I STUT STUT TER TER)
```

3. Modify function DOUBLE in exercise 3 in such a way as to double all the atoms in a list, whatever their depth.

4. Write a predicate function, with two arguments, which returns the value *true* if the first argument occurs within the list which appears as the second argument. Below are a few examples of applications of this function:

```
(SAME 'B '(A B C D))           → (B C D)
(SAME 'Z '(A B C D))           → ()
(SAME 3 '(1 2 3 4))            → (3 4)
```

5. Modify the preceding function so that the element the function looks for can be at any depth and be of any type (i.e. a list or an atom).

6. Write a function which groups together the successive elements of two lists. Here are a few examples showing what the function should do:

```
(GR '(A B C) '(1 2 3))              → ((A 1)(B 2)(C 3))
(GR '(M N O) '(13 14 15 16))        → ((M 13)(N 14)(O 15) 16)
(GR '(M N O P) '(13 14 15))         → ((M 13)(N 14)(O 15) P)
```

7. What does the following function do:

```
(DEFUN FOO (X)
  (IF (NULL X) NIL (APPEND (FOOBAR X X) (FOO (CDR X)))))
```

with function FOOBAR defined as follows:

```
(DEFUN FOOBAR (X Y)
  (IF (NULL Y) ()
    (CONS X (FOOBAR X (CDR Y)))))
```

8. What does the following function do:

```
(DEFUN F (X Y)
        (IF X (CONS (CAR X) (F Y (CDR X))) Y))
```

9. What does the following function do:

```
(DEFUN BAR (X)
        (IF (NULL (CDR X)) X
            (CONS (CAR (BAR (CDR X)))
                  (BAR (CONS (CAR X)
                        (BAR (CDR (BAR (CDR X)))))))))
```

7 ARITHMETIC

7.1 BASIC ARITHMETIC FUNCTIONS

Though LISP is essentially a language for symbolic and therefore non-numeric programming, it does allow for programming numeric problems. Here we shall only consider calculations with integer numbers between -2^{15} and $2^{15}-1$.[1] Let us first of all consider standard arithmetic functions:

$$(1+\ n) \rightarrow n+1$$

Function 1+ adds the value 1 to its argument. It is an *incrementation* function.

$$(1-\ n) \rightarrow n-1$$

Function 1− subtracts the value 1 from its argument. It is a *decrementation* function.

Note 1

Note that 1+ and 1− are *names* of functions which, like any LISP function, are written in prefix form, i.e. before the argument. 1+ or 1− are therefore names like CAR or CDR, and the sign + or − must be written in such a way that it immediately follows the letter 1 (with no space between the 1 and the + or −).

Note 2

Note also that neither function 1+ nor function 1−, nor for that matter any other basic arithmetic function, modifies the values of its arguments: it simply calculates a value.

Here are the addition (+), subtraction (−), multiplication (*) and division (/) functions[2]:

$$(+\ n1\ n2) \rightarrow n1 + n2$$
$$(-\ n1\ n2) \rightarrow n1 - n2$$
$$(*\ n1\ n2) \rightarrow n1 * n2$$
$$(/\ n1\ n2) \rightarrow n1\ /\ n2$$

All these arithmetic operations are generic, i.e. the arguments can be of any numeric type. If all the arguments are integers, the result will also be an integer, otherwise the

[1] In classical LISPs the minimum and maximum values of numbers depend on the particular machine on which you are running LISP. Nevertheless, Common-LISP automatically converts to another representation if the numbers become too large or too small.

[2] In general the functions +, − and * will take any number of arguments. Before beginning to write long programmes on a machine, you should first test these various arithmetic functions.

result will depend on the types of the argument. The exception to this rule is the division function /: in some LISP implementations this division is an *integer division*, which means that the result of dividing $n1$ by $n2$ is the largest integer x such that

$$(x^{*}n2) \leq n1$$

which is normally written: $\lfloor n1/n2 \rfloor$.

In LE_LISP, on the other hand, the result of a division, using function /, of an integer n by an integer n leads to an interrupt if n is not a multiple of m. To obtain integer division, as we have just described, using two integer arguments, LE_LISP provides the function QUO.

In Common-LISP the result of the division of integers is either an integer or a fraction. This preserves the correction in expressions where the result of the division is used for other calculations.

There is also a function to calculate the *remainder after integer division*, which is written REM.

$$\text{(REM } n1\ n2) \rightarrow n1 - ((\lfloor n1/n2 \rfloor) * n2)$$

Here are a few numeric examples with their result in Le_LISP and Common-LISP:

Common-LISP:		Le_LISP	Common-LISP
(1+ 1024)	→	1025	1025
(1- 0)	→	−1	−1
(1- −345)	→	−346	−346
(+ 17 9)	→	26	26
(+ 1 −1)	→	0	0
(− 10 24)	→	−14	−14
(*6 4)	→	24	24
(/ 1024 2)	→	512	512
(/ 17 7)	→	2	17/7[3]
(REM 1024 2)	→	0	0
(REM 17 7)	→	3	3
(+ (* 3 3)(* 4 4))	→	25	25
(− 111(* 4 (/ 111 4)))	→	3	0
(/ (+10 90)(− (* 5 6) 10))	→	5	5

These standard functions allow us to write a large number of useful functions. Here, for example, is a function which calculates the *square* of its argument:

```
(DEFUN SQUARE (N) (* N N))
```

[3] To get in Common-LISP the integer result of an integer division, one has to use the function FLOOR on the resulting fraction. For example: (FLOOR (/ 17 7)) → 2.

and here are a few further simple and very useful arithmetic functions, starting with two functions which calculate the *square* of the *square* of their argument:

```
(DEFUN SQUARE-OF-SQUARE (N) (SQUARE (SQUARE N)))
```

or, more simply:

```
(DEFUN SQUARE-OF-SQUARE (N) (* N (* N (* N N))))
```

or, even more simply:

```
(DEFUN SQUARE-OF-SQUARE (N) (* N N N N))
```

here is a function which calculates the expression (A + B)/(A − B):

```
(DEFUN SUM-DIFF (N M) (/ (+ N M) (− N M)))
```

What does the following function do?

```
(DEFUN FOO (L1 L2)
       (+ (* (CAR L1)(CAR L2))(* (CADR LA) (CADR L2))))
```

7.2 NUMERIC PREDICATES

In order to be able to write more interesting arithmetic functions, we first need to study some numeric predicates. Here are the most important:

First a predicate which tests whether its argument is equal to 0

(ZEROP *n*)	→ T if $n = 0$ → NIL if $n \neq 0$

the predicate = should be read *is numerically equal to* and it tests the equality of two numbers:

(= *arg1 arg2*)	→ T if *arg1* = *arg2* → NIL if they are different

predicate > is read *greater than*:

(> *n1 n2*)	→ T or *n1* if *n1* is greater than *n2* → NIL if *n1* is less than or equal to *n2*

predicate < is read *less than*:

(< *n1 n2*)	→ T or *n1* if *n1* is less than *n2* → NIL if *n1* is greater than or equal to *n2*

predicate >= is read *greater than or equal to*:

(>= *n1 n2*)
→ T or *n1* if *n1* is greater than or equal to *n2*
→ NIL if *n1* is less than *n2*

predicate <= is read *less than or equal to*:

(<= *n1 n2*)
→ T or *n1* if *n1* is less than or equal to *n2*
→ NIL if *n1* is greater than *n2*

Here are two further predicates, one called EVENP which tests whether its argument is an even number, and another called ODDP which tests whether its argument is an odd number. They are defined as follows:

(EVENP *n*)
→ T or *n* if *n* is even
→ NIL if *n* is odd

(ODDP *n*)
→ T or *n* if on is odd
→ NIL if *n* is even

Writing these two functions for Common-LISP should present no problem:

```
(DEFUN ODDP (N) (IF (= (REM N 2) 1) T NIL))

(DEFUN EVENP (N) (IF (ZEROP (REM N 2)) T NIL))
```

7.3 NUMERIC PROGRAMMING

To become a little more familiar with numeric algorithms, we shall, in the rest of this chapter, give a few of the best known. Let us start with the ubiquitous function FACTORIAL. The factorial of a number *n*, for those who are unfamiliar with the concept, is the product of all the positive integers less than or equal to *n*. Consequently we have the following definitions:

Factorial (n) = n*(n − 1) * (n − 2) *. . .*3*2*1
Factorial (0) = 1

In mathematical textbooks, you will often find the following recurrent definition:

factorial (0) = 1
factorial (n) = n*factorial (n − 1)

This definition gives us directly the recursive algorithm we need to construct. Let us translate it into a LISP function:

```
(DEFUN FACTORIAL (N)
        (IF (ZEROP N) 1
                (* N (FACTORIAL (1- N)))))
```

This, give or take a few textual transpositions, is the same as the mathematical definition. Just for the fun of it, here are a few calls and their results:

```
(FACTORIAL 0)    →   1
(FACTORIAL 1)    →   1
(FACTORIAL 2)    →   2
(FACTORIAL 3)    →   6
(FACTORIAL 4)    →  24
(FACTORIAL 5)    → 120
```

This function grows very quickly.

Did you know, by the way, that there is only one number for which the equation:

$$\text{factorial}(n) = \text{factorial}(a) * \text{factorial}(b)$$

has a non-trivial solution? It is:

$$\text{factorial}(10) = \text{factorial}(7) * \text{factorial}(6)$$

In his time, Euclid developed an algorithm to find the *highest common factor*, abbreviated to HCF, of two numbers.[4] His algorithm is as follows:

to calculate the HCF of m and n proceed as follows:
if m is greater than n calculate the HCF of n and m
else, if the remainder on dividing n by m is equal to 0
  if $m = 1$ then m and n have no common factors
    else m is the HCF of n and n
  else we have the relation:
    $\text{HCF}(m, n) = \text{HCF}(\text{remainder } \lfloor (n/m) \rfloor, m)$

Clearly, recurrent definitions are common in mathematics. Here is a translation of Euclid's algorithm into a recursive LISP function:

```
(DEFUN HCF (M N)
        (LET ((X (HCF1 M N)))
                (IF X X (LIST M 'AND N 'ARE 'MUTUALLY 'PRIME))))
(DEFUN HCF1 (M N) (COND
        ((> M N) (HCF1 N M))
        ((= (REM N M) 0)
                (IF (= M 1) () M))
        (T (HCF1 (REM N M) M)) ))
```

[4] There is considerable information on Euclid's algorithm and other numerical algorithms in the second volume of D. E. Knuth's monumental work. This book is such a source of information that it is almost obligatory reading. Here are the references: D. E. Knuth, *The Art of Computer Programming*; the first volume is entitled *Fundamental Algorithms*, the second *Semi-Numerical Algorithms* and the third *Searching and Sorting*. The whole is published by Addison-Wesley.

LET is not a function you know yet. LET is used here to store the value of the call (HCF1 M N) somewhere, in order to be able to access it later. We could have written function HCF as follows:

```
(DEFUN PHCF (M N)
      (IF (HCF1 M N) (HCF1 M N)
          (LIST M 'AND N 'ARE 'MUTUALLY 'PRIME)))
```

But if the function is written in this way, the call to (HCF1 M N) would be calculated *twice*: once for the IF predicate, to determine whether there is an HCF of M and N and again to work out the value of the call. Which, to say the least, would be far from elegant. What we needed was a means to bind this intermediate value to a variable. This is one of the commonest uses of the LET function.

LET is a way of defining a *nameless* function. Here is its syntactic definition:

(LET ($item_1$ $item_2$ . . . $item_n$) *function-body*)

with each of the *items* defined as follows:

(*variable value*)

The semantics, i.e. the meaning, of the function LET is as follows:

> On entry to LET the various variables are bound to the values given in the *variable-value* pairs in the LET variable list. Next, the *function-body* of the LET is evaluated using these bindings. The value of a LET is the value of the last expression evaluated (as during the evaluation of normal user functions). It is therefore both a definition of a function (though a function without a name) and its call.

Let us now return to the function HCF: on entry to the LET, variable X is bound to the value of the call to the auxiliary function HCF1. This value may be a number, if the arguments have an HCF, or NIL if the arguments are mutually prime. Depending on the value of X, the function either returns the value of the HCF or a list indicating the two arguments are mutually prime. Here are some calls and their values:

```
? (HCF 256 1024)
= 256

? (HCF 3456 1868)
= 4

? (HCF 567 445)
= (567 AND 445 ARE MUTUALLY PRIME

? (HCF 565 445)
= 5
```

```
? (HCF 729 756)
= 27

? (HCF 34560 18369)
= 9
```

Here is another numeric function: INTEGER-ROOT, a function to compute the integer root of a number. The integer root x of a number n is defined as the largest integer x such that

$$(x \star x) \leqslant n$$

where $\leqslant$ means *less than or equal to.*

```
(DEFUN INTEGER-ROOT (N)
       (INTEGER-ROOT-AUX 1 0 (1- N)))

(DEFUN INTEGER-ROOT-AUX (P K N)
      (IF (< N 0) K
          (INTEGER-ROOT-AUX (+ P 2) (1+ K) (- N (+ 2 P)))))
```

The function INTEGER-ROOT is definitely very short: two lines! Actually it doesn't really compute anything. All it does is to prepare arguments for the function INTEGER-ROOT-AUX, to which all the computation is delegated. INTEGER-ROOT-AUX is, as its name indicates, an auxiliary function. It *only* serves the INTEGER-ROOT function and will never be called from any other function.

In cases like this, i.e. when we construct an auxiliary function which will only be called from *one* other place, or from *one* other function, it is good programming practice to write this auxiliary function in such a way that it is *local* to the calling function. *Local to the calling function* means: only known to that function, and outside that calling function it is unknown. This has two advantages: first, one is assured that no error occurs because of the calling of this local function from somewhere else; second, one can choose an arbitrary name for this auxiliary function, without worrying that this name may be already in use somewhere else, since one can define local functions using *identical* names at different places without destroying the other ones.

Here is our Common-LISP definition of INTEGER-ROOT using local functions:

```
(DEFUN INTEGER-ROOT (N)
       (LABELS ((AUXILIARY (P K N)
                           (IF (< N 0) K
                               (AUXILIARY (+ P 2) (1+ K) (- N (+ 2 P))))))
               (AUXILIARY 1 0 N)))
```

The LABELS function permits us to define local functions. These local functions are only known inside the call to the LABELS function. Here is the syntactic definition of LABELS:

(LABELS ((*name1 variable-list function-body*)
(*name2 variable-list function-body*)
. . . .
(*name-n variable-list function-body*))
labels-body)

which means: the first argument of function LABELS is a list of function definitions. Each function definition is of the form

(*name variable-list function-body*)

where *name* is the name of the function, *variable-list* is the list of variables of the function, and *function-body* is the body of the function, that is a sequence of expressions to evaluate.

Labels-body is a sequence of expressions to evaluate. The different functions *name1*, *name2*, . . . *name-n* are local to the *labels-body* and can only be called from within it (or from within the function itself, in the case of recursive definitions). In our example, the function AUXILIARY is only known inside the LABELS construct of the function INTEGER-ROOT.

In Le_LISP, or other LISPs of that family, local functions are constructed using a special form of LET. The LET function is, as we have already pointed out, both the definition and the call to a function.

The special form of the LET function which permits us to call recursively the function defined by LET, is called LETN and has as an additional first argument the name which the function defined through the LET may temporarily have. LETN is a named LET and is defined as follows:

(LETN *name* (*item1 item2* . . . *item-n*) *function-body*)

This form has the same behaviour as the LET function, *except* that the symbol *name* is associated with the expression itself, within the LETN. This makes it possible to call LET recursively. Naturally, it is only used when one wants to construct a local function.

To see how this functions, let us have a look at the Le_LISP definition of function INTEGER-ROOT.

```
(DEFUN INTEGER-ROOT (N)
       (LETN SELF ((P 1)(K 0)(N (1- N))
          (IF (< N 0) K
             (SELF (+ P 2) (1+ K) (- N (+ 2 P)))))))
```

The function SELF refers back to the function constructed using the LETN function. This SELF function is only known inside the body of the named LET (the LETN function) and is not defined outside.

What is the *algorithm* of INTEGER-ROOT? Why does this function give us the integer-root of a number? To show that it really works, here are a few examples:

```
(INTEGER-ROOT 9)        →    3
(INTEGER-ROOT 10)       →    3
(INTEGER-ROOT 8)        →    2
(INTEGER-ROOT 49)       →    7
(INTEGER-ROOT 50)       →    7
(INTEGER-ROOT 12059)    →  109
```

Here a few more examples of numeric functions. First the Le_LISP function DEC-BIN which translates a decimal number into a list of zeros and ones (i.e. into a list representing a binary number):

```
(DE DEC-BIN (N)
   (IF (> N 0)
      (APPEND (DEC-BIN (/ N 2)) (APPEND (REM N 2) NIL))
      NIL))
```

For this function, the definition of function APPEND (cf. section 6.2) has been slightly modified. What modification has been made? Why is it necessary?

Here are a few examples of calls:

```
? (DEC-BIN 5)
= (1 0 1)

? (DEC-BIN 1023)
= (1 1 1 1 1 1 1 1 1 1)

? (DEC-BIN 1024)
= (1 0 0 0 0 0 0 0 0 0 0)

? (DEC-BIN 555)
= (1 0 0 0 1 0 1 0 1 1)
```

If you have not yet found the modification that needs to be made to function APPEND, remember that it concatenates two *lists*, and that here the first argument may be an atom, as in the call (APPEND (REM N 2) NIL).

Here then is the new APPEND function:

```
(DEFUN APPEND (ELEMENT LIST) (COND
        ((NULL ELEMENT) LIST)
        ((ATOM ELEMENT) (CONS ELEMENT LIST))
        (T(CONS (CAR ELEMENT) (APPEND (CDR ELEMENT) LIST)))))
```

In order to write this function BIN-DEC in Common-LISP we have to remember that the division function '/' does *not* truncate the result to its integer part when applied to integers. Nevertheless, that is exactly why this function works in Le_LISP.

Here we have to use the function FLOOR of Common-LISP. FLOOR is a very interesting Common-LISP function: it is a function which returns *two* values. Here is an example:

(FLOOR 10 3) → 3 1

the first value is the largest integer less than or equal to the quotient (the same number as the result of Le_LISP's integer division), the second value is the difference between the first argument, here 10, and the first value times the second argument, short: it is the remainder.

Here are some other examples of the use of FLOOR:

```
(FLOOR 17 7)            →  2 3
(FLOOR (/ 17 7) 1)      →  2 3/7
(FLOOR (/ 17 7))        →  2 3/7
(FLOOR 1023 17)         → 60 3
(FLOOR 483 23)          → 21 0
```

Note that the second argument defaults to 1.

Multiple values – functions returning more than one value – are a very special thing in Common-LISP. We won't enter here into their details and just note that if such functions are used as arguments of other functions not expecting multiple values, only the first value will be used. For example:

(1+ (FLOOR 483 23)) → 22

The second value is lost here.

After this diversion, here is finally the function DEC-BIN in Common-LISP:

```
(DEFUN DEC-BIN (N)
   IF (> N 0)
      (APPEND (DEC-BIN (FLOOR N 2)) (CONS (REM N 2)NIL))
      NIL))
```

Naturally, if we have a function for translating decimal numbers into binary numbers, we also need the inverse function: function BIN-DEC which translates binary numbers into decimal numbers.

```
(DEFUN BIN-DEC (N)
    (LABELS ((AUXILIARY (N RES)
                 (IF N
                     (AUXILIARY (CDR N) (+ (* RES 2) (CAR N)))
                     RES)))
           (AUXILIARY N 0)))
```

Here are a few examples of its use:[5]

```
? (BIN-DEC '(1 0 1))
= 5

? (BIN-DEC '(1 0 0 0 0 0 0 0 0 0 0))
= 1024

? (BIN-DEC '(1 0 1 0 1 0 1 0 1 0 1))
= 1365

? (BIN-DEC '(1 1 1))
= 7
```

Study these functions carefully. Try them by hand, and try running them on the machine.

7.4 EXERCISES

1. Write a function, called NTH, with two arguments, a number n and a list l, which returns the n-th element of list l.

   Here are a few possible calls to this function:

   ```
   (NTH 2 '(DO RE MI FA))            → RE
   (NTH 3 '(DO RE MI FA))            → MI
   (NTH 1 '((WELL WELL) AHA AHA))    → (WELL WELL)
   ```

2. Write a function to convert decimal numbers into octal or hexadecimal numbers.

3. Write a function to translate octal numbers into decimal numbers, and another to translate hexadecimal numbers into decimal numbers.

---

[5] Remember that the program BIN-DEC will be written in Le_LISP as follows:

```
(DE BIN-DEC (N)
    (LETN AUXILIARY ((N N) (RES 0))
        (IF N (AUXILIARY (CDR N) (+ (* RES 2) (CAR N)))
            RES))
```

4. In 1220, a great mathematical problem in the region of Pisa was the following: if a pair of rabbits has two offspring every month, one male and one female, and if after a month the new pair can also produce two offspring of opposite sex, how many rabbits are there after n months: Leonardo of Pisa (usually known as Fibonacci) established the following recurrent rule:

```
fibonacci (0)  = 1
fibonacci (1)  = 1
fibonacci (n)  = fibonacci (n - 1) + fibonacci (n - 2)
```

Write a function which will calculate fibonacci numbers for any positive argument n.

5. What does the following function do:

```
(DEFUN WHAT (N)
        (LABELS ((SELF (RES N)
                (IF (> N 0)
                    (SELF (+ (* RES 10) (REM N 10)) (FLOOR N 10))
                    RES)))
                (SELF 0 N)))
```

6. Function ACKERMANN is defined as follows:

```
ack(0,n)        = n + 1
ack(m, 0)       = ack (m-1, 1)
ack(m, n)       = ack (m-1, ack(m, n - 1))
```

Write this function in LISP. Do you have any idea what it does?

7. Write a function LENGTH which calculates the number of elements in a list. Here are a few examples of calls to it:

```
(LENGTH '(LITTLE BY LITTLE))              → 3
(LENGTH '(((ART) ADJ) NOUN VERB))         → 3
(LENGTH '((DO RE MI)))                    → 1
(LENGTH ())                               → 0
```

8. Write a function which calculates the number of atoms in a list. Here are some examples:

```
(NBATOM '(LITTLE BY LITTLE))              → 3
(NBATOM '(((ART) ADJ) NOUN VERB))         → 4
(NBATOM '((DO RE MI)))                    → 3
(NBATOM '())                              → 0
```

8 P-Lists

So far we have only used *symbols*, i.e. *literal atoms*, either as *names*, or as *parameters* in user functions. In the latter case, symbols may have *values*: the values to which they are bound when calls are made to the functions for which they are parameters. We call the value of the binding a *C-value* or *cell-value*. LISP also provides associative database mechanisms through its *P-lists* or *property lists*. It is these lists with which we shall be concerned in this chapter.

8.1 SYMBOLS

Let us first consider the internal representation of symbols. One symbol can be distinguished from another principally by three characteristics: its *name* (its external representation), its *value* (the value to which it is bound) and its *P-list*. The value of a symbol is given by the function SYMBOL-VALUE. The P-list of a symbol is given by the function SYMBOL-PLIST. Consequently, the call

```
(SYMBOL-VALUE 'XYZZY)
```

calculates the C-value of symbol XYZZY,[1] and the call

```
(SYMBOL-PLIST 'XYZZY)
```

returns the P-list of the same symbol.[2] Note that in Le_LISP the function PLIST may optionally take a second argument. In this case the list passed as second argument becomes the new P-list of the symbol appearing as first argument.

If symbol XYZZY is associated with the numerical value 100, the LISP expression:

```
(1+ (SYMBOL-VALUE 'XYZZY))
```

has the same value as the expression

```
(1+ XYZZY)
```

[1] XYZZY is a magic word well known to those who play *adventure*. If you do not yet know it, try it out, and you will be amazed by its instantaneous effect. You could also try the magic word PLUGH.

[2] In Le_LISP the function SYMBOL-VALUE is called SYMEVAL and the function SYMBOL-PLIST becomes PLIST in Le_LISP.

In both cases the value would be 101. This property can be extremely useful from time to time.

Here is how a symbol can be represented internally:

C-value *its value*
P-list *its list of properties*
P-name the *name* of the *atom*

Initially, the C-value of a symbol is the value UNDEFINED, indicating that the symbol is not bound to a value and any request for access to its value will cause an *error*: unbound variable. The P-list is initialized to the value NIL. Naturally, the *P-name* – or *print-name* – will be the sequence of characters giving the atom's name. Here is the internal representation of the symbol XYZZY we met above.

100 the value of the atom is 100
NIL its P-list is equal to NIL
XYZZY its print name is XYZZY

8.2 ACCESS TO P-LISTS

A symbol will often have more than one value. For example, to implement a database of family relations, it may be necessary to represent somewhere that PETER has a father called JOHN, a mother called JULIE, and a son called GERALD. This means that the value of property *father* of PETER is JOHN, that the value of property *mother* of PETER is JULIE and that the value of property *son* of PETER is GERALD. Obviously, the values associated with PETER are multiple and depend on the particular characteristic being queried. LISP provides three functions affecting P-lists to implement such relations: PUTPROP, GET and REMPROP.

Let us now look at an example of the use of these functions, and then at their definitions. Here is a list of LISP instructions to implement the family relations for PETER.

```
(PUTPROP 'PETER 'FATHER 'JOHN)

(PUTPROP 'PETER 'MOTHER 'JULIE)

(PUTPROP 'PETER 'SON 'GERALD)
```

These constructions have the effect of putting the atom JOHN into the P-list of symbol PETER under the indicator FATHER, the atom JULIE under the indicator MOTHER and the atom GERALD under the indicator SON.

The P-list of a symbol is therefore a series of indicators and of values associated with those indicators:

$$(indicator_1\ value_1\ indicator_2\ value_2\ \ldots\ indicator_n\ value_n)$$

After having executed the calls to PUTPROP just mentioned, the P-list of the symbol PETER is of the following form:

```
(FATHER JOHN MOTHER JULIE SON GERALD)
```

P-lists may be seen as tables of correspondences associated with symbols. Hence the P-list given above will be represented as:

PETER

Indicator	MOTHER	SON	FATHER
Value	JULIE	GERALD	JOHN

Function GET is used to find out the value associated with an indicator. For example, to discover PETER's father, it is only necessary to ask:

```
(GET 'PETER 'FATHER)
```

which returns the atom JOHN as value.

Here is a definition of function PUTPROP

(PUTPROP *symbol indicator value*) → symbol

which has the following effect:

> PUT places value *value*, under indicator *indicator*, in the P-list of symbol *symbol*. If the indicator *indicator* already exists, the previously associated value will be overwritten. *All* the arguments are evaluated (this is why we quoted them). The value returned by function PUTPROP is the value of its first argument, i.e. the value of *symbol*.

In MACLISP, the LISP dialect that runs under Multics, and other operating systems, the function PUTPROP returns *value* as value. In Le_LISP, the arguments for PUTPROP appear in the following order:

(PUTPROP *symbol value indicator*).

Here is the definition of function GET (which is called GETPROP in Le_LISP):

(GET *symbol indicator*) → *value*

GET returns the value associated with indicator *indicator* in the P-list of symbol *symbol*. If the indicator is not in the P-list of *symbol*, GET returns the value NIL. *There is no way, therefore, of distinguishing between a value NIL associated with an indicator, and the absence of the indicator altogether.*

This is why the function GET permits an optional third argument in Common-LISP. This third argument indicates the value of GET in the case that the searched property doesn't exist. For example, the call

```
(GET 'PETER 'AGE 'UNKNOWN)
```

returns the value associated to the property AGE on the P-list of the symbol PETER if this property exists otherwise, instead of returning NIL, this call to GET returns the symbol UNKNOWN. This permits us to distinguish easily between the absence of a property and the value NIL of a property.

Actually we cheated a little: the function PUTPROP exists in all dialects of LISP *except* in Common-LISP. There, in order to use it, we should first define this function as follows:

```
(DEFUN PUTPROP (SYMBOL INDICATOR VALUE)
       (SETF (GET SYMBOL INDICATOR) VALUE)
       SYMBOL)
```

The function SETF is a very general assignment function which we will examine in detail later. For the moment just note that – in Common-LISP – a property value is given by

(SETF (GET *symbol indicator*) *new-value*)

So, without the function PUTPROP just defined, the construction of the property list of the symbol PETER could as well have been done with the following three calls:

```
(SETF (GET 'PETER 'FATHER) 'JOHN)
(SETF (GET 'PETER 'MOTHER) 'JULIE)
(SETF (GET 'PETER 'SON) 'GERALD)
```

Function REMPROP removes a property from a P-list. Consequently, after the call:

(REMPROP 'PETER 'SON)

the pair SON – GERALD will be removed from the P-list of atom PETER, and the instruction

(GET 'PETER 'SON)

will return the value NIL, reflecting the absence of indicator SON.

(REMPROP *symbol indicator*) → *symbol*

REMPROP removes the indicator *indicator*, as well as its associated value, from symbol *symbol*. REMPROP returns the symbol given as first argument.

8.3 EXERCISES

1. Use the functions PUTPROP, GET and REMPROP to translate the following sentences into LISP:

    a. Peter is Guy's father.
    b. Peter also has a child called Mary.
    c. Peter also has a child called James.
    d. Peter is of the male sex.
    e. Mary is of the female sex.
    f. Guy and James are of the male sex.
    g. Judy is of the same sex as Mary.
    h. Judy is 22 years old.
    i. Anne is 40 years old.
    j. Sarah's age is the sum of Judy's and Anne's.

2. (Project) Assume that every book in a library is represented by an atom and that the whole library is represented by a list containing these atoms. Assume also that each book has a property called *title*, a property called *author* and a third called *keywords*. Build up a small bibliographic database, with interrogation functions for this database containing these three properties. Naturally, if there are several possible replies to a request, the interrogation function must return a list of all possible replies. Consequently, if you ask for a book or books by Karl Kraus, for example, and if your library has the two titles *Die letzten Tage der Menschheit* and *Von Pest und Presse*, the function must return the list ((DIE LETZTEN TAGE DER MENSCHHEIT) (VON PEST UND PRESSE)). To give yourself an idea of the problems associated with such databases, take a look at a book on *automatic document retrieval*.

3. We give below a small family tree, with arrows going from parents towards children. Represent the simple family relations (father and mother) and sexes on the P-lists

of the atoms for the names of different people. Next, write a program which will discover relations that are not directly represented, such as brother, sister, son, daughter, uncle, aunt, cousin, grandfather and grandmother relations.

Finally, define a function ANCESTORS, which returns a list of all the known ancestors of a person.

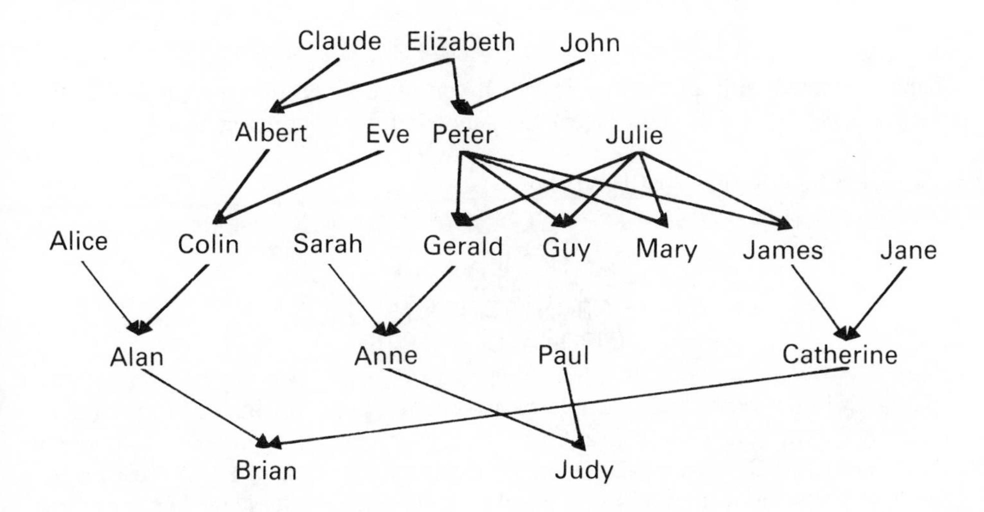

9 Memo Functions

Chapter 7 ended with an exercise in which you were asked to write the FIBONACCI function. Using the definition given, the simplest form of this function is:

```
(DEFUN FIBONACCI (N)
        (COND
             ((= N 0) 1)
             ((= N 1) 1)
             (T (+ (FIBONACCI (1- N))
                   (FIBONACCI (- N 2)) )) ))
```

FIBONACCI is a very good example of a recursive function, but it is unfortunately extremely inefficient. This is clear if we look at it closely: in order to calculate the value of FIBONACCI for a number n, it calculates the value of FIBONACCI $n - 1$, then $n - 2$, etc, until it ends with n equal to 1. The same values are then recalculated for the second recursive call (with the exception of FIBONACCI $n - 1$). Here is a trace of a call to this function:

```
?(FIBONACCI 4)
   → FIBONACCI: (4)
     → FIBONACCI: (3)
       → FIBONACCI: (2)
         → FIBONACCI: (1)
         ← FIBONACCI = 1
         → FIBONACCI: (0)
         ← FIBONACCI = 1
       ← FIBONACCI = 2
       → FIBONACCI: (1)
       ← FIBONACCI = 1
     ← FIBONACCI = 3 ;end of first recursive call;
     → FIBONACCI: (2)
       → FIBONACCI: (1)
       ← FIBONACCI = 1
       → FIBONACCI: (0)
       ← FIBONACCI = 1
     ← FIBONACCI = 2 ;end of second recursive call;
   ← FIBONACCI = 5
= 5
```

Intuitively, it seems horrifying that the machine should have to recalculate all these values which it has already calculated before. *Memo functions* make intensive use of P-lists to store values already calculated in the course of calls, making it possible in this way to *remember* or *store* the function value. Here is how the FIBONACCI function could be defined as a memo-function:

```
(DEFUN FIB (N)
   (COND
      ((= N 0) 1)
      ((= N 1) 1)
      ((GET 'FIB N)) ;this is where the saving is made;
      (T (PUTPROP 'FIB N
                  (+ (FIB (1 - N)) (FIB (-N 2))))
         (GET 'FIB N))))
```

Each time, before it carries out a recursive calculation of the value of a call, this function looks first in its P-list to see whether it does not already have a value for this call, and if so, FIB returns that value. Otherwise, FIB calculates the value, puts it in the P-list (for later use) and also returns it. Compare the trace for the calculation of FIB 4 with the trace obtained for the FIBONACCI function:

```
? (FIB 4)
   → FIB: (4)
     → FIB: (3)
       → FIB: (2)
         → FIB: (1)
         ← FIB = 1
         → FIB: (0)
         ← FIB = 1
       ← FIB = 2
       → FIB: (1)
       ← FIB = 1
     ← FIB = 3                        ;end of first recursive call;
     → FIB: (2)
     ← FIB = 2                        ;end of second recursive call;
   ← FIB = 5
   = 5
```

Now look at the trace for the subsequent call to FIB 5:

```
? (FIB 5)
   → FIB: (5)
     → FIB: (4)                       ;first recursive call;
     ← FIB = 5
     → FIB: (3)                       ;second recursive call;
   ← FIB = 3
← FIB = 8
= 8
```

When the function encounters the recursive call to (FIB 4) it can immediately return the value 5, thanks to the value stored in the P-list for FIB. Here is the P-list after these calls:

```
(5 8 4 5 3 3 2 2)
```

where value 8 appears under indicator 5, value 5 under indicator 4, value 3 under indicator 3 and value 2 under indicator 2. Consequently, for each call to the function which has already been calculated once, the result is immediately found by consulting the P-list, and no recalculation takes place.

Think of this technique, if you have a *very* recursive function which you are using frequently within a program.

It is possible to improve this FIB function still further by putting the values of FIB for 0 and 1 in the P-list, making it unnecessary to include the tests:

```
((= N 0) 1)
((= N 1) 1)
```

This speeds up the calculation.

Here is the FIB function modified in this way, first for Le_LISP:

```
(DE FIB (N)
    (IF (<= N 1) 1                                    ;to avoid problems
        (UNLESS (GET 'FIB 0)
              (PUT 'FIB 0 1)
              (PUT 'FIB 1 1);
         (LETN SELF ((N N)) (COND
              ((GET 'FIB N))
         (T (PUT 'FIB N
                   (+(SELF (1- N)) (SELF (- N 2))))
            (GET 'FIB N))))))
```

and then for Common-LISP:

```
(DEFUN FIB (N)
    (LABELS ((AUX (N) (COND
                        ((GET 'FIB N))
                        (T (SETF (GET 'FIB N)
                                 (+ (AUX (1- N)) (AUX (- N 2))))))))
           (COND
               ((<= N 1) 1)
               (T (UNLESS (GET 'FIB 0)
                       (PUTPROP 'FIB 0 1)
                       (PUTPROP 'FIB 1 1))
                  (AUX N)))))
```

In this versions we have used a new function: UNLESS. Like function IF, this is a choice function. Its definition is:

(UNLESS *test* $action_1$ $action_2$. . .$action_n$)	→ evaluates $action_1$ to $action_n$ and returns $action_n$ as value if *test* = NIL → otherwise it returns NIL

the call

(UNLESS *test* $action_1$ $action_1$ . . . $action_n$)

is therefore identical to a call to the following IF function:

(IF *test* NIL $action_1$ $action_2$ . . . $action_n$)

and is simply a simplification of it.

There is also a function WHEN which is the inverse of UNLESS:

(WHEN *test* $action_1$ $action_2$. . .$action_n$)	→ evaluate $action_1$ to $action_n$ and return $action_n$ as value if *test* $\neq$ NIL → NIL if *test* = NIL

The call

(WHEN *test* $action_1$ $action_2$ . . . $action_n$)

can therefore be written as:

(UNLESS (NULL *test*) $action_1$ $action_2$ $action_n$)

or again:

(IF (NULL *test*) () $action_1$ $action_2$ . . . $action_n$)

You can therefore choose from amongst IF, COND, UNLESS and WHEN, to write choice functions.

9.1 EXERCISES

1. Write function FACTORIAL as a memo-function. Compare the execution time of factorial with the use of a P-list with that of the function without using the P-list.

2. Look at the following program:

```
(DEFUN FOO (X) (COND
          ((< X 0) (- 0 X))
          (T (COND ((ZEROP X) 0)
                   (T (* X -1 ))))))
```

a. Simplify this program, taking into consideration only its syntactic structure, so that it uses only one COND.

b. Given the meaning of the various arithmetic functions, simplify the program so that it no longer contains any COND functions at all.

3. Write a small algebraic simplifier. This simplifier should be able to simplify LISP arithmetic expressions (i.e. well parenthesized algebraic expressions) and be aware, at least, of the following simplifications:

(+ e 0)	→ e
(+ 0 e)	→ e
(* e 1)	→ e
(* 1 e)	→ e
(* e 0)	→ 0
(* 0 e)	→ 0
(+ *number1 number2*)	→ *number1* + *number2*
(* *number1 number2*)	→ *number1* * *number2*

For example, the program should be able to simplify

(+ (* x 0) (* 10 (+ y 0)))

to

(* 10 y)

10 INPUT–OUTPUT (PART ONE)

We have only considered one way of communicating with our machine: evaluation either of a function call or of a variable. The only values that we can give to functions are those given in the call, and the only values appearing on the terminal screen are those resulting from an evaluation requested at the user–machine interaction level, i.e. at *top level*. We may however want to display intermediate results, or other information, or even to write programs allowing dialogue with the machine. Functions allowing printing or reading in the middle of evaluation (or of execution) of a program are called *input–output* functions. This chapter will be concerned with the definition and use of such functions.

10.1 THE GENERAL PRINT FUNCTION PRINT

PRINT is the LISP function used to print values. This function is the second we have met which is used not only to obtain the value it returns after evaluating its argument, but also because of the *side effect* that it causes.[1] Naturally, this side effect is the printing of the value of its arguments.

Here is the definition of function PRINT:

(PRINT *arg*) → arg

plus: printing the values of the argument

PRINT is therefore a function with any number of arguments. On being called, the argument is evaluated *and* printed on the standard output peripheral (display screen, line printer, disc file or magnetic tape, depending on the context in which you are working).[2] The value of the argument is returned as the value of the function, just as all other functions return a usable value following computation.

Here are two simple examples of the use of this function:

---

[1] The other two functions with *side effects* that we already know are functions SETF and REMPROP, which return a value and also produce the side effect of modifying a P-list.

[2] We shall see later how a standard peripheral may be changed.

```
? (PRINT 4)
  4                                        ;printing;
= 4                                        ;value returned;
? (PRINT (CONS 'WELL '(IT WORKS)))
    (WELL IT WORKS)
                                           ;printing;
= (WELL IT WORKS)                          ;value returned;
```

Before actually printing, PRINT automatically moves to the next line.

In Le_LISP, and other dialects of LISP, PRINT accepts an arbitrary number of arguments, which are all evaluated and printed. The value returned is the value of the last argument. Such, in Le_LISP, one may write:

```
(PRINT 'A 'B 'C 'D)
```

which will print:

```
ABCD
```

and return D as its value.

Here is a little function which prints all the elements in a list:

```
(DEFUN PRINT-ALL (L) (COND
            ((NULL L) '(NO MORE))
            (T (PRINT (CAR L))
               (PRINT-ALL (CDR L)))))
```

and here are a few calls to the function:

```
?    (PRINT-ALL '(DO RE MI FA SOL))
      DO
      RE
      MI
      FA
      SOL
=    (NO MORE)

?    (PRINT-ALL '(ONE TWO THREE))
      ONE
      TWO
      THREE
=    (NO MORE)
```

After each print operation, the machine executes a *carriage return* and a *line feed* (i.e. the cursor moves to the beginning of the next line).

To give a less trivial use of function PRINT, here is a solution to the *Hanoi Towers* problem.[3] The problem is as follows: three discs have been stacked on a needle, in order of decreasing size, as follows:

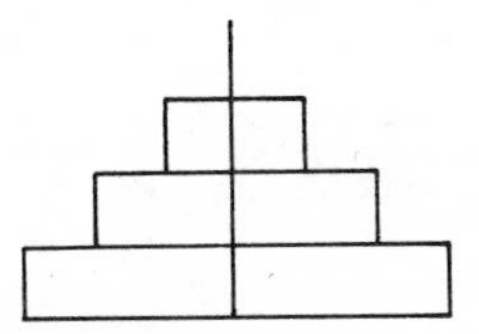

Close by there are two other needles. The discs have to be moved from the first needle to the second, and the third may be used as a temporary storage place. The following constraints have to be respected at all times:

1. Only one disc can be moved at a time.

2. No disc must ever be on a smaller disc.

Here is the solution for three discs:

1. Initial state

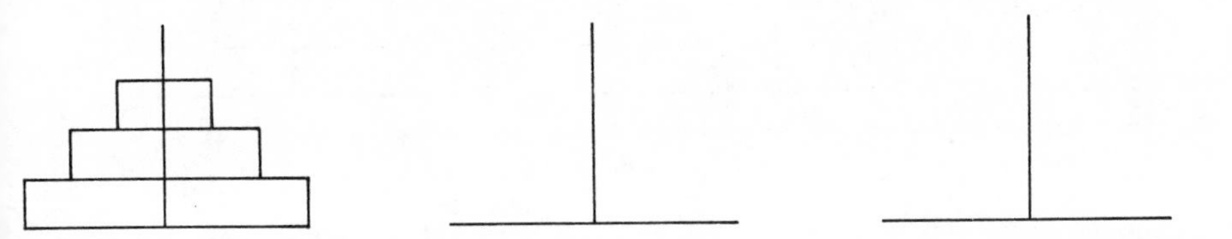

2. First movement

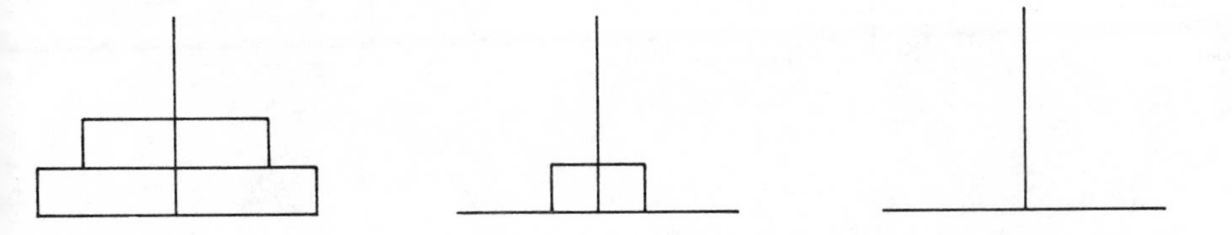

[3] According to an ancient myth, a group of monks in an oriental temple spend their time transferring 64 golden discs from one needle to another, according to the rules given above. The existence of the universe as we know it would, it was believed, be completed once the task was finished.

Incidentally, what would the total age of the universe be if it took one second to move each disc?

3. Second movement

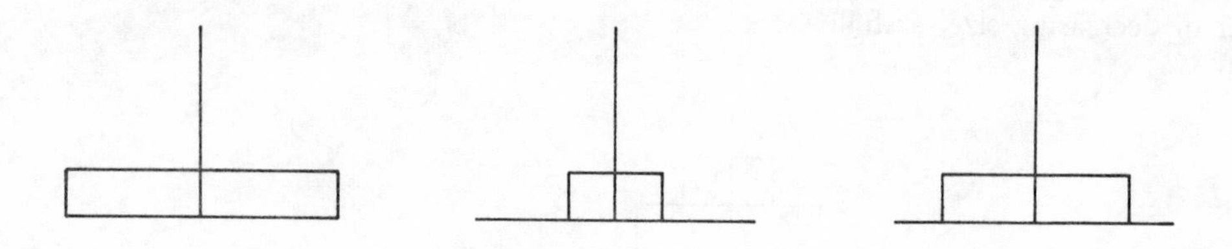

4. Third movement

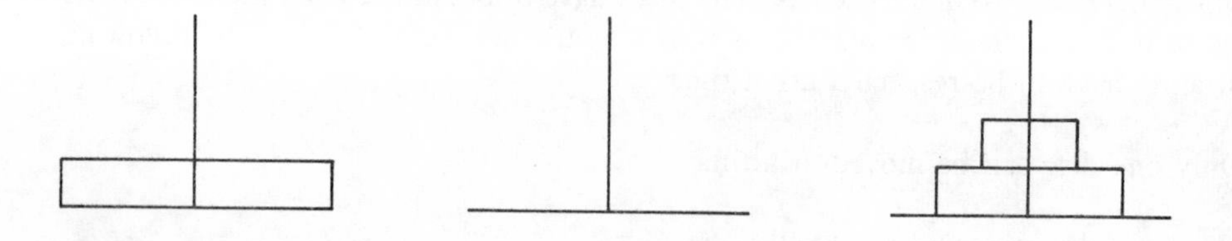

5. Fourth movement

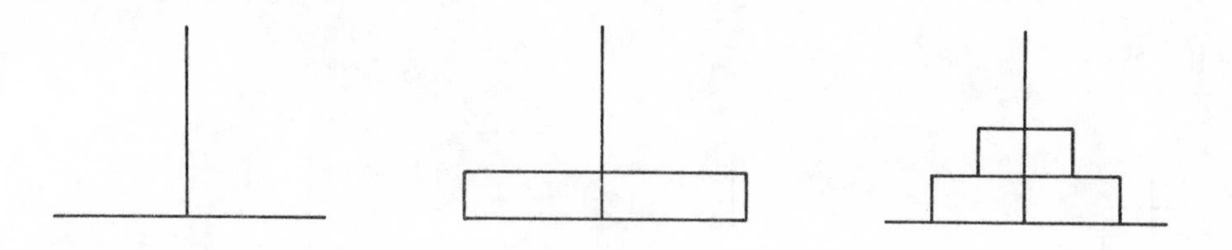

6. Fifth movement

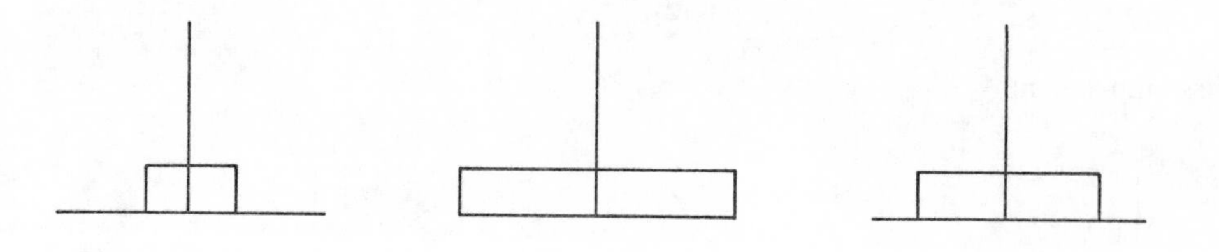

7. Sixth movement

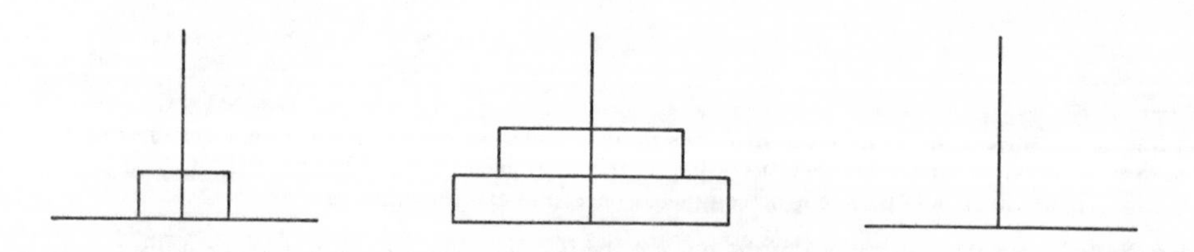

8. Final state

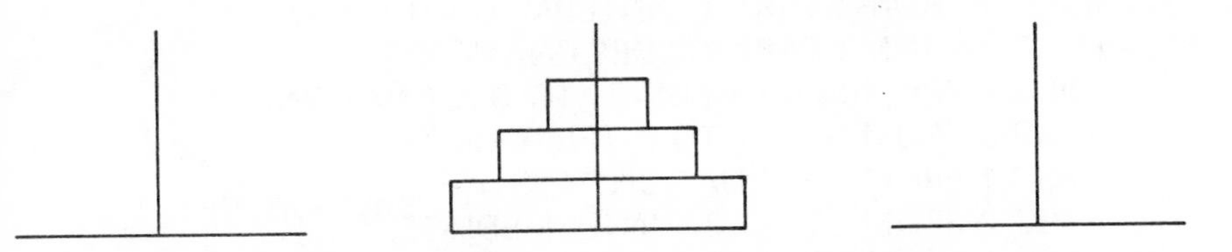

In the fourth figure the condition of the needle being used as an intermediate stage is similar to the initial state. The difference between the two states is that the biggest disk is not included in the configuration on the intermediate needle. It is this observation that gives us the solution to the problem. Note that the first three movements have simply built a new *Hanoi Tower* on the intermediate needle, with one disc less than on the original tower. As the *destination* needle is free at that moment, it is now possible to move the biggest disk on to it, from the *starting needle* where it is now alone. We then start the operation again, with the needles' roles interchanged: what was originally the intermediate needle is now the starting needle (as all the discs to move are on it). On the other hand, the original starting needle is now the intermediate needle.

A generalization of this observation gives us the following algorithm:

```
(DEFUN HANOI (N START DESTINATION INTERMEDIATE)
     (COND
        ((= N 0) '(FINISHED AT LAST))
        (T (HANOI (1- N) START INTERMEDIATE DESTINATION)
           (PRINT (LIST 'DISC N 'FROM START 'TO DESTINATION))
           (HANOI (1- N) INTERMEDIATE DESTINATION START))))
```

and here is an example of a few calls:

```
?    (HANOI 2 'START 'DESTINATION 'INTERMEDIATE)
     (DISC 1 FROM START TO INTERMEDIATE)
     (DISC 2 FROM START TO DESTINATION)
     (DISC 1 FROM INTERMEDIATE TO DESTINATION)
=    (FINISHED AT LAST)

?    (HANOI 3 'START 'DESTINATION 'INTERMEDIATE)
     (DISC 1 FROM START TO DESTINATION)
     (DISC 2 FROM START TO INTERMEDIATE)
     (DISC 1 FROM DESTINATION TO INTERMEDIATE)
     (DISC 3 FROM START TO DESTINATION)
     (DISC 1 FROM INTERMEDIATE TO START)
     (DISC 2 FROM INTERMEDIATE TO DESTINATION)
     (DISC 1 FROM START TO DESTINATION)
=    (FINISHED AT LAST)
```

```
?    (HANOI 4 'START 'DESTINATION 'INTERMEDIATE)
     (DISC 1 FROM START TO INTERMEDIATE)
     (DISC 2 FROM START TO DESTINATION)
     (DISC 1 FROM INTERMEDIATE TO DESTINATION)
     (DISC 3 FROM START TO INTERMEDIATE)
     (DISC 1 FROM DESTINATION TO START)
     (DISC 2 FROM START TO INTERMEDIATE)
     (DISC 1 FROM START TO INTERMEDIATE)
     (DISC 4 FROM START TO DESTINATION)
     (DISC 1 FROM INTERMEDIATE TO DESTINATION)
     (DISC 2 FROM INTERMEDIATE TO START)
     (DISC 1 FROM DESTINATION TO START)
     (DISC 3 FROM INTERMEDIATE TO DESTINATION)
     (DISC 1 FROM START TO INTERMEDIATE)
     (DISC 2 FROM START TO DESTINATION)
     (DISC 1 FROM INTERMEDIATE TO DESTINATION)
=    (FINISHED AT LAST)
```

Analyse this function carefully and build up a call tree for it: it is interesting, and not only for its use of function PRINT.

10.2 CHARACTER STRINGS

So far we have only considered three types of LISP objects: *lists*, *numbers* and *atoms*. In order to make output more aesthetically pleasing, we may need to print special characters such as *spaces*, *points* or any other character which has so far been inaccessible because it normally acts as a *separator* in LISP. *Character strings* are the LISP objects allowing access to these characters. No check is carried out on the syntactic role they usually play, if they are included in a *string* (the shortened form of the expression '*character string*').

Syntactically, any string is written in quotation marks ("):

character string ::= "*sequence of characters*"

where a *sequence of characters* may be made up of any characters.

Here are a few examples of strange but wholly correct strings:

```
"aBcDeFgHiJk"
"(((((((((("
".<>.<>."
"$%_& ' ()_@"
```

Strings are LISP objects of the *atom* type. Predicate ATOM applied to a string, or to a variable whose value is a string, therefore returns the value *true*, i.e. T, as in the following example:

```
(ATOM "aBc.cBa" → T
```

To distinguish between strings and other atoms, LISP uses predicate STRINGP defined as follows:

```
                     →T in Common-LISP, string e in Le_LISP, if
                       argument e is a character string.
(STRINGP e)
                     →NIL in all other cases.
```

Here are a few examples of applications of this predicate, first in Common-LISP

```
(STRINGP "LISP is beautiful")                → T
(STRINGP 'CUBE1)                             → NIL
(STRINGP (CAR '("hey" is that true?)))       → T
(STRINGP '("well" "well" "a sausage"))       → NIL
```

and now in le-LISP:

```
(STRINGP "LISP is beautiful")                → "LISP is beautiful"
(STRINGP 'CUBE1)                             → ()
(STRINGP (CAR '("hey" is that true?)))       → "hey"
(STRINGP '("well" "well" "a sausage"))       → ()
```

NOTE

Note that character strings should not be *quoted*. They are *constants* like atom T and atom NIL. They are atoms which are their own values.

Here are a few examples of uses of function PRINT with strings:

```
?   (PRINT 1)                                ;as usual;
    1
=   1

?   (PRINT "          ")                     ;with spaces;
    "          "
=   "          "

?   (PRINT "........")                       ;do you see the point?;
    "........"
=   "........"
```

For the moment, you should just remember that each time we have to print special characters, we shall use character strings.

10.3 FUNCTIONS PRIN1, TERPRI AND PRINCH

Printing or, more generally, any output normally takes place in two stages: first a copy of what you want to print is made in a special zone of computer memory, a zone generally called a *buffer*, and then the content of this memory zone or buffer is sent to the output peripheral: the buffer is *flushed*.

The PRINT function combines the two stages: it fills the output buffer with information and then sends it to the peripheral. But it is also possible to separate the stages using functions PRIN1 in Common-LISP or function PRIN in Le_LISP, followed by TERPRI. PRIN1 is a function which fills, or as we usually say *edits*, the buffer. This function goes on to actual printing only once the buffer is completely full. While it is not full, nothing appears on the terminal (if the terminal is the standard output peripheral). To flush the buffer, function TERPRI has to be called. It empties the buffer and repositions an invisible pointer at the start of the buffer. If the buffer is empty when the function is called, TERPRI prints nothing and moves to the next line.

To make things clearer, assume we have the following buffer:

```
|                                                        |
 ↑
```

with the pointer, represented here by the small arrow, at the beginning of the buffer. If we evaluate the call

```
(PRIN1 'HELLO)          ; in Common-LISP

(PRIN 'HELLO)           ;in Le_LISP
```

the effect is as follows:

```
|HELLO                                                   |
      ↑
```

The sequence of characters making up the atom HELLO has added to the start of the buffer and the pointer has moved on until it is pointing to the first free location. The calls

```
(PRIN1" Josephine")(PRIN1 "...")        ;in Common-LISP;
(PRIN " Josephine")(PRIN  "...")           ;in Le_LISP;
```

will modify the state of the buffer as follows:

HELLO Josephine ...

↑

No actual printing has yet taken place. All that PRIN1 does is to add material to the output buffer. For printing to start, the TERPRI function has to be explicitly called:

(TERPRI)

which would lead in Le_LISP to the appearance of the following line:

HELLO Josephine . . .

and in Common-LISP the line

HELLO" Josephine"" . . ."

and the buffer would return to its initial state:

↑

i.e. the buffer is empty and its pointer is pointing at the first location.

Each time function PRIN1 is used, all calls to that function must be followed by at least one call to function TERPRI.

Formally, PRIN1 is defined as follows:

(PRIN1 *arg*) → *arg*

PLUS: *filling of output buffer with the value of the argument.*

PRIN1 evaluates the argument, adds it to the output buffer and returns its value. PRIN1 does not lead to actual printing unless the buffer is completely full, and no further elements can be added to it.

(TERPRI) → NIL

In Le-LISP TERPRI has an optional argument which, if it is present, must be a number. TERPRI prints the output buffer. If the buffer is empty the operation has the same effect as a line feed. If an argument *n* is present, function TERPRI flushes the buffer (i.e. it prints the contents of the buffer) and carries out *n − 1 additional line feeds*. TERPRI returns the value of its argument, if there is one, and NIL otherwise.

Functions PRINT and PRIN1 print the objects in such a way so as LISP could read it; for example, strings are printed with surrounding double quotes. The Le_LISP function PRINCH, an acronym for PRIN CHaracter, has two arguments: a character to print and a number indicating how many times the character is to be printed. Successive appearances of this character are not separated by spaces, but printed one immediately after the other. The value returned by a call to function PRINCH is the character given as first argument. Here are a few examples of the function:

```
? (PRINCH " "5)
          =           ;this prints character " " 5 times and returns it
                      as value. Here the value is printed on the same
                      line as the 5 occurrences of the character, since
                      function PRINCH does not see to printing, but
                      merely fills the buffer. Printing is carried out by
                      the top level of LISP;
? (PRINCH "."10)
  ..........=.        ;the final point is the one returned as value;
? (PRINCH 1)          ;if there is no second argument, LISP assumes
  1 = 1               a default value of 1
```

Common-LISP doesn't have this function PRINCH. But one of the very powerful characteristics of LISP, of any LISP dialect whatsoever, is the possibility to define one's own functions which may then be used like standard LISP functions. So, let us define the PRINCH function.

```
(DEFUN PRINCH (STRING N)
   (COND
      ((ZEROP N) STRING)
      (T(PRINC STRING)(PRINCH STRING (1- N)) )))
```

This function uses the Common-LISP standard function PRINC. PRINC is identical to PRIN1, except that strings are printed without the surrounding double quotes. Here are two examples of the use of PRINC

```
  (PRINC ".")
  .
= "."
  (PRINC "hello World")
  hello World
= "hello World"
```

To give an example of the use of these functions, the following is the definition of a *formatting* function, as in FORTRAN. This function has two arguments, a list of elements to be printed and a list of columns giving the tab values at which the successive elements should be printed.[4]

[4] Function PLENGTH which calculates the number of characters in the name of an atom will be examined in the next section.

```
(DEFUN FORMATTING (ELEMENTS COLUMNS)
   (TERPRI)                     ;for an empty line;
   (FORMAT-AUX 0 ELEMENTS COLUMNS))

(DEFUN PRINT-FIRST-ELEMENT (N ELE COL)
   (COND
        ((= N COL) (PRING ELE) N)
        (T (PRINC " ")
        (PRINT-FIRST-ELEMENT (1 + N) ELE COL))))

(DEFUN FORMAT-AUX (N ELEMENTS COLUMNS)
        (COND
          ((NULL ELEMENTS) NIL)                          ;finished;
          ((NULL COLUMNS) NIL)                           ;finished;
          (T (FORMAT-AUX
                    (+ (PRINT-FIRST-ELEMENT N (CAR ELEMENTS)
                              (CAR COLUMNS))
                       (PLENGTH (CAR ELEMENTS)))
                    (CDR ELEMENTS)
                    (CDR COLUMNS)))))
```

The construction:

```
(COND
     ((NULL ELEMENTS) NIL)
     ((NULL COLUMNS) NIL)

                    . . .
```

indicates that computation will be complete as soon as list ELEMENTS or list COLUMNS is empty. To test whether one at least of a set of tests is true, LISP provides function OR defined as follows:

(OR *test*$_1$ *test*$_2$. . .*test*$_n$)
→the result of the first *test* which does not evaluate to NIL
→NIL if all the tests *test*$_i$ evaluate to NIL

We can therefore rewrite our test in the form:

```
(COND
     ((OR (NULL ELEMENTS) (NULL COLUMNS)) NIL))
```

Function OR works exactly like a *logical or*.

We also have functions AND which corresponds to a *logical and* which checks whether all the tests in a set of tests are true. Here is its definition:

(AND $test_1$ $test_2$. . .$test_n$)

→ $test_n$ if none of the $test_i$ evaluates to NIL

→ NIL if at least one $test_i$ evaluates to NIL

There is also an auxiliary function F1 whose arguments appear in a list containing sub-lists, each of which gives the elements to print in the format shown by the second argument. Here it is:

```
(DEFUN F1 (LIST-OF-ELEMENTS COLUMNS)(COND
          ((NULL LIST-OF-ELEMENTS) (TERPRI))
          (T(FORMATTING (CAR LIST-OF-ELEMENTS) COLUMNS)
               (F1 (CDR LIST-OF-ELEMENTS) COLUMNS))))
```

and here, finally, are a few examples of print operations that this function produces:

```
(F1 '((ENGLISH FRENCH GERMAN)
    (  --------------------  ------------------------------  ------------------------------ )
    (COMPUTER ORDINATEUR COMPUTER)
    (STACK PILE KELLER)
    (MEMORY MEMOIRE SPEICHER)
    (DISPLAY ECRAN BILDSCHIRM)
    (KEYBOARD CLAVIER TASTATUR))
  '(5 25 45))

→

ENGLISH                       FRENCH                        GERMAN
------------------------------  ------------------------------  ----------------------------------

COMPUTER                      ORDINATEUR                    COMPUTER
STACK                         PILE                          KELLER
MEMORY                        MEMOIRE                       SPEICHER
DISPLAY                       ECRAN                         BILDSCHIRM
KEYBOARD                      CLAVIER                       TASTATUR

= NIL
```

The next example is a table of monthly prices for two French telecommunication networks:

```
(F1 '((COSTS TELEPHONE LINK CADUCEE TRANSPAC)
     (CONNECTION 4800F 3200F 3200F 4000F)
     (SUBSCRIPTION 189F 4003F 3130F 1810F)
     (USE 13536F 0F 4060F 117F))
     '(4 20 30 40 50))

→
```

```
COSTS           TELEPHONE   LINK    CADUCEE   TRANSPAC
TRANSPAC
LINK            4800F       3200F   3200F     4000F
SUBSCRIPTION    189F        4003F   3130F     1810F
USE             13536F      0F      4060F     117F

= NIL
```

Actually, it is not necessary to really write the function FORMATTING in such a complicated way: using the Common-LISP standard function FORMAT we can do the same much more easily. Using this function, here is the new code for the function FORMATTING:

```
(DEFUN FORMATTING (ELEMENTS COLUMNS) (COND
                 ((NULL ELEMENTS) (TERPRI)) ;the end!;
                 (T (FORMAT T "~VT~A" (CAR COLUMNS)
                                    (CAR ELEMENTS))
                   (FORMATTING
                           (CDR ELEMENTS) (CDR COLUMNS)))))
```

The function FORMAT is Common-LISP's special function to help outputting in nice ways without writing complicated programs. Actually, each call to FORMAT is a little program, and the programming language used is composed by the signs in between the double quotes. Here is a more formal definition of FORMAT:

(FORMAT T *a-control-string a-sequence-of-arguments*)

The first argument T indicates where the printing has to take place; we will come back to this later in this book. The second argument, *a-control-string*, indicates how the arguments composing the *a-sequence-of-arguments* are to be printed.

The simplest way to call format is with a control-string without any special character ~ (tilde) in it. In this case, the control string is printed as it is. Here is a snapshot of an interaction with Common-LISP using FORMAT in this easy way:

```
? (FORMAT T "Hello World")
Hello World
=
? (FORMAT T "aha @:[]{}#")
aha @:[]{}#
=
?
```

Called with a control-string without the tilde in it, FORMAT behaves almost like the function PRINC with the same string as argument. The difference is the returned value: PRINC returns the string given as an argument, FORMAT returns NIL.

Normally, when printing, one wants to print the content of variables, arbitrary LISP objects. To do this, for each object to print, one has to include ~A in the control-string and then, after the control string give the arguments in the order one wants to print them. For example:

```
(FORMAT T "~A~A~A" 1 2 3)
```

prints

```
123
```

and

```
(FORMAT T "~A ~A rA" 1 2 3) ;note the spaces!!;
```

prints

```
1 2 3
```

The meaning of this is: FORMAT takes the control-string, replaces each occurrence of ~A by the corresponding value given as argument after the control-string, leaves all the other characters in place, and then outputs the obtained string. Here are some more examples of the use of FORMAT:

```
? (FORMAT T "~A is not equal to ~A" 1 2)
1 is not equal to 2
=
? (FORMAT T "~2A~2A~2A" 1 2 3)
 1 2 3
=
? (FORMAT T "~2A~2A~2A" 1 '(A B C) 3)
 1(A B C D) 3         ;note the spacing;
=
? (FORMAT T "~2,2,2A~2,2,2A~2,2,2A" 1 '(A B C) 3)
  1 (a b c) 3
=
```

The control argument ~A is the short form of ~*i,j,k,l*A which causes the argument to be padded at the right with at least *k* (default 0) copies of the padding character *l* (default is space), then padding characters are inserted *j* (default 1) at a time, until the output is at least of length *i* (default 0). In the last example above, we gave values for *i*, *j* and *k*, and took the default padding character.

Following is a list of possible control-string directives:

~D the argument must be a number which will be printed as a decimal integer. ~nD uses a column width of *n*; spaces are inserted on the left if the number requires less than *n* columns for its digits and sign.

~O this is like ~D but the number is printed in octal instead of decimal.

~X this is like ~D but in base 16

~S this is like ~A, but the argument is printed rather like with PRIN1 than like with PRINC.

~% Outputs a *newline*. -n% outputs *n* newlines.

~~ Outputs a tilde; ~n~ outputs *n* tilde characters.

~T The T stands for *tabulation*. ~i,jT spaces, if possible, over to column *i*; otherwise, it moves to columns *i* +*k***j* for the smallest positive integer *k* for which this is possible.

In place of a numeric argument to such a directive, you can put the letter V, which takes an argument from *a-sequence-of-arguments* as an argument to the directive, that is what we did in the function FORMATTING.

10.4 ACCESS TO SYMBOL-NAMES

The LISP atom, like the physical atom, is indivisible in appearance only. The LISP atom equivalent of the elementary particles making up a physical atom are the characters making up the LISP atom's name. In Le_LISP, these characters are made accessible by function EXPLODECH and the formation of an atom by fusing elementary particles is made possible by function IMPLODECH.

Let us take just one step at a time, and start by considering function EXPLODECH, defined as follows:

(EXPLODECH *atom*) → a list of characters corresponding to the sequence of characters making up the external and therefore printable form of the value of *atom*.

Function EXPLODECH has one argument, which it evaluates; if the argument's value is an atom, it returns the list of characters making up the atom; if the value is a list, it returns NIL. Here are a few examples of the use of this function:

```
? (EXPLODECH 'ATOM)
= (A T O M)
```

```
? (EXPLODECH 123)
= (1 2 3)

? (EXPLODECH (CAR '(HE IS LEAVING)))
= (H E)

? (CDR (EXPLODECH 'AVERYLONGWORD))))
= (V E R Y L O N G W O R D)

? (LENGTH (EXPLODECH 'HEAVENS-MY-HUSBAND))
= 18
```

The Le_LISP function EXPLODECH is more general: it takes any LISP expression as argument and returns the sequence of characters making up that expression. For example, if you call:

```
(EXPLODECH '(CAR ' (A B)))
```

the result is the list:

```
(|(| C A R |'||(| A B|)||)|)
```

Special characters are enclosed in two "|" signs. In Le_LISP, this is a way of quoting characters, without taking their special role into account.

Determining the length of the list resulting from a call to function EXPLODECH is the same as calculating the number of characters making up the atom's P-name. This may be a very useful piece of information if a pretty-printing function is being built up by hand, as in the case of function FORMATTING, since it makes it possible to calculate the exact position of the pointer in the output buffer. As we need to know the length of the P-name of an atom fairly often, Le_LISP provides a special function PLENGTH, which can be defined as follows:

```
(DEFUN PLENGTH (ELT)
        (IF (ATOM ELT) (LENGTH (EXPLODECH ELT) )0))
```

Here are a few calls to function PLENGTH:

```
(PLENGTH 'HOP)                  → 3
(PLENGTH 'MARYLIN)              → 7
(PLENGTH (CAR '(A B C)))        → 1
```

Of course, if we can break the name of an atom down into a sequence of characters, we would also like to be able to take a sequence of characters and form an atom from it. This operation, the inverse of function EXPLODECH, is carried out by function IMPLODECH.

IMPLODECH takes a list of characters as argument and returns as result an expression which is made up of those characters. Consequently, we have the following relation:

(IMPLODECH (EXPLODECH list-of-characters)) → *expression*

Let us examine a few examples of the use of IMPLODECH:

```
? (IMPLODECH '(A T O M))
= ATOM

? (IMPLODECH '(C U R I O U S)) = CURIOUS

? (IMPLODECH (APPEND (EXPLODECH 'BIRTH) (EXPLODECH 'DAY)))
= BIRTHDAY

? (IMPLODECH (EXPLODECH 'IDENTITY))
= IDENTITY
```

IMPLODECH is a very convenient way of creating new atoms, which can be particularly useful from time to time, especially in natural language applications.

The functions IMPLODECH, EXPLODECH and PLENGTH do not exist in Common-LISP. In other dialects they exist mainly to be able to write functions for nice printing of LISP objects. Given that Common-LISP has the FORMAT function and a very nice set of string handling functions (which we will see later), these functions are not necessary. Nevertheless, let us define them in Common-LISP. Through it we will discover some important characteristics and functions of Common-LISP.

Here is the EXPLODECH function:

```
(DEFUN EXPLODECH (SYMBOL)
   (EXPL (STRING SYMBOL) 0 (LENGTH (STRING SYMBOL))))

(DEFUN EXPL (SEQ COUNT N) (COND
       ((= COUNT N) ())
       (T (CONS (ELT SEQ COUNT)(EXPL SEQ (1+ COUNT) N))) ))
```

Let us explain: Common-LISP distinguishes between different objects such as numbers, lists, strings, symbols (these are non-numeric atoms), characters and arrays. For these different objects exist special functions permitting to access or to modify them. For example, the function LENGTH calculates the length of a list and the function REVERSE computes a copy of its argument, a list, where the order of the elements is inversed. Sometimes it may be interesting to know the number of characters composing a string of characters or the number of elements of a vector (a vector is a one-dimensional array). All this counting of characters or elements defines the *length* of the corresponding object. That is why Common-LISP introduces an additional data-type: the *sequence*. A sequence is composed of lists, vectors or strings. By the way, a string is nothing else but a vector of characters. Or, more precisely: lists, vectors and strings are considered to be sequences, such that if we say a function applies to sequences, we mean it is applicable to lists, strings, and vectors. Functions applying to sequences are REVERSE, DELETE and LENGTH.

In the definition of the function EXPLODECH above, we used the function LENGTH on the argument L (which is a symbol, an atom) after having it converted to a string thanks to the function STRING.

Here is the definition of the function STRING:

(STRING *object*)	→ a string derived from *object*. If *object* is a string, it is returned, if it is a symbol, its print name is returned, if it is a character, the string containing that one character is returned. Passing any other object to STRING may result in an error.

And here are three examples of its use:

```
(STRING 'AHA)                  → "AHA"
(STRING "So what is That?")    → "So what is That?"
(STRING   '#\c)                → "c"
```

In the last example the argument is a character: this is Common-LISP's way to enter a character-object. If you are very, very much interested in knowing more about characters, just go ahead to the next chapter where we will see more about different characters in Common-LISP and Le_LISP, but, probably, it is better just to follow the 'wait and see' strategy.

Just to verify that it works, let's look at some of the sequence handling functions:

```
(REVERSE '(A (B C) D E))      →  (E D (B C) A)
(DELETE 'A' (A B A C A))      →  (B C)
(DELETE '#\A "ABACA")         →  "BC"
(REVERSE "aBcDe")             →  "eDcBa"
(LENGTH "ABCDE")              →  5
(LENGTH '(A (B C) D E))       →  4
```

There is one more sequence handling function we should have a look at: the function CONCATENATE. This function concatenates two lists, two vectors or two strings. Here is its definition:

(CONCATENATE *result-type sequence-1 sequence-2*)

a new sequence of type *result-type* containing all the elements of *sequence-1* and *sequence-2* in order.

Here are some examples of the use of CONCATENATE:

```
(CONCATENATE 'LIST '(A B) '(C D))        → (A B C D)
(CONCATENATE 'LIST '(A B) '((C D) E))    → (A B (C D) E)
```

working on lists, the function CONCATENATE behaves exactly like the function APPEND which we defined in Chapter 6. Let us look at some more examples:

```
(CONCATENATE 'STRING "YOU " "HERE?")            → "YOU HERE?"
(CONCATENATE 'LIST '(A B) "ABCD")               → (A B #\A #\B #\C
                                                   #\D) !!!
(CONCATENATE 'STRING "AIR" (STRING 'PLANE))     → "AIRPLANE"
```

To understand our definition of EXPLODECH, we need to know still one more function: the function ELT. ELT is, like REVERSE, CONCATENATE, DELETE and LENGTH, a sequence handling function. It is defined as follows:

(ELT *sequence index*) → the *index* element of *sequence*. If *sequence* is a string, ELT returns a character, if it is a list, it returns the corresponding element of the list. Note that the index of the first element of a sequence is 0 (zero).

Here are some examples of ELT:

```
(ELT '(A B C D) 3)                    → C
(ELT "oh, that is all?" 9)            → #\i
(ELT '(1 2 3 4) (ELT '(4 3 2 1) 3))   → 1
```

After all this, please look back at our Common-LISP definition of the function EXPLODECH; try to understand the underlying algorithm (it shouldn't be too hard!).

Actually, there is another very useful sequence-handling function: the function REDUCE. REDUCE is a function with two arguments: the first argument is itself a function and the second a sequence. The effect of evaluating a call to REDUCE is to apply the function given as the first argument to the first two elements of the sequence given as second argument; the result thus obtained is then given again to the function (as a first argument) and the next element constitutes the second argument to the function. This goes on until all the elements of the sequence have been given as an argument to the function.

This sounds very complicated. It is not! Look at the following example:

```
(REDUCE #'+ '(1 2 3 4 5)) → 15
```

which is obtained by calculating:

```
(+ (+ (+ (+ 1 2) 3) 4) 5)
```

This is evidently a very useful function for recursively applying a function of two arguments to a sequence of an arbitrary number of arguments. For example, to apply our function APPEND of chapter 6 to an arbitrary number of lists, it is sufficient to pack all these lists in one big list and then to REDUCE APPEND on this list. Thus

```
(REDUCE 'APPEND '((A B) (C D) (E F) (G H) (I J) (K L))
```

yields as a result the concatenation of all the sub-lists of the second argument, i.e.: the list

```
(A B C D E F G H I J K L)
```

Well, equipped with all these new functions we can define a Common-LISP function IMPLODECH and PLENGTH. Here they are:

```
(DEFUN IMPLODECH (LIST)
   (IF (< (LENGTH LIST) 2) (STRING (CAR L))
      (REDUCE #'STRING-APPEND L)))
```

with STRING-APPEND defined as:

```
(DEFUN STRING-APPEND (X Y)
         (CONCATENATE 'STRING (STRING X) (STRING Y)))
```

Finally, the function PLENGTH is just the length of a string:

```
(DEFUN PLENGTH (X) (LENGTH (STRING X)))
```

The rest of this section will discuss a small program to conjugate regular French verbs of the first group, i.e. regular verbs ending in *er*, *ir*, or *re*.

;the function CREATE-WORD *creates a new word from a root and an ending;*

```
(DEFUN CREATE-WORD (ROOT ENDING)
   (IF (NULL ENDING) ROOT
      (IMPLODECH (APPEND (EXPLODECH ROOT)
                 (EXPLODECH ENDING)))))
```

;the function ROOT *finds the root of a verb, by simply removing its last two characters;*

```
(DEFUN ROOT (VERB)
   (IMPLODECH (REVERSE (CDR (CDR
                          (REVERSE (EXPLODECH VERB)))))))
```

;the function TYPE *returns the atom made up of the last two characters of the atom given as argument;*

```
(DEFUN TYPE (VERB)
      (LET ((AUX (REVERSE (EXPLODECH VERB))))
          (IMPLODECH (CONS (CADR AUX)
                        (CONS (CAR AUX) ())))))
```

;CONJUGATE *is the function the user calls. It prepares arguments for the auxiliary function* CONJ1;

```
(DEFUN CONJUGATE (VERB)
   (TERPRI)                                   ;to make things prettier;
      (CONJ1 (ROOT VERB)                      ;find the root;
         (TYPE VERB)                          ;and its type;
         '(JE TU IL ELLE NOUS VOUS ILS ELLES)))
```

;CONJ1 *does all the work; it first prepares the list of endings and stores them in* ENDINGS; *next, the auxiliary function DO-IT prints the conjugation for each of the pronouns;*

```
(DEFUN CONJ1 (ROOT TYPE PRONOUNS)
   (LABELS ((DO-IT (PRONOUNS ROOT ENDINGS) (COND
               ((NULL PRONOUNS) "that's it")
               (T (PRIN1 (CAR PRONOUNS))
                  (PRIN1 (CREATE-WORD ROOT (CAR ENDINGS)))
                  (TERPRI)
                  (DO-IT (CDR PRONOUNS) (CDR ENDINGS))))))
            (DO-IT PRONOUNS ROOT
                  (COND
                     ((EQ TYPE 'ER) '(E ES E E ONS EZ ENT ENT))
                     ((EQ TYPE 'RE) '(S S T T ONS EZ ENT ENT))
                     (T '(IS IS IT IT ISSONS ISSEZ
                              ISSENT ISSENT))))))
```

Study this program carefully: it is the first program in this introduction to LISP which is made up of a combination of several functions. To understand how it operates, load it onto the machine and let it run. Next, once it works, start to modify it, by adding, for example, verbs in *oir*.

But let us first look at a few calls, to ensure that it really works and that it does what we wanted.

```
?   (CONJUGATE 'PARLER)        ;first a verb in 'er';

    JE PARLE
    TU PARLES
    IL PARLE
    ELLE PARLE
    NOUS PARLONS
    VOUS PARLEZ
    ILS PARLENT
    ELLES PARLENT
=   that's it

?   (CONJUGATE 'FINIR)         ;next a verb in 'ir';

    JE FINIS
    TU FINIS
    IL FINIT
    ELLE FINIT
    NOUS FINISSONS
    VOUS FINISSEZ
    ILS FINISSENT
    ELLES FINISSENT
=   that's it

?   (CONJUGATE 'ROMPRE)        ;and finally a verb in 're';

    JE ROMPS
    TU ROMPS
    IL ROMPT
    ELLE ROMPT
    NOUS ROMPONS
    VOUS ROMPEZ
    ILS ROMPENT
    ELLES ROMPENT
=   that's it
```

This program is, of course, extremely limited: regular verbs give correct results. For all others, anything might happen, depending on their endings. In the exercises below I suggest a few improvements, or extensions, to be made to this program.

10.5 EXERCISES

1. Add a further argument to function CONJUGATE, showing the tenses for which you want the conjugation of the verb given as first argument. So, for example, the call

```
(CONJUGATE 'FINIR 'PRESENT-SUBJUNCTIVE)
```

should print:

```
QUE JE FINISSE
QUE TU FINISSES
QU'IL FINISSE
QU'ELLE FINISSE
QUE NOUS FINISSIONS
QUE VOUS FINISSIEZ
QU'ILS FINISSENT
QU'ELLES FINISSENT
```

Modify the program so that it can conjugate the following tenses:

a. Present (e.g. je parle)
b. Imperfect (e.g. je parlais)
c. Future (e.g. je parlerai)
d. Past historic (e.g. je parlai)
e. Conditional (e.g. je parlerais)
f. Present subjunctive (e.g. que je parle)
g. Imperfect subjunctive (e.g. que je parlasse)
h. Perfect (e.g. j'ai parle)
i. Pluperfect (e.g. j'avais parlé)
j. Past anterior (e.g. j'eus parlé)
k. Past conditional – 1 (e.g. j'aurais parlé)
l. Past conditional – 2 (e.g. j'eusse parlé)

It will not be enough simply to include a call to function PRINT in function CONJ1. We should therefore introduce a function VERBPRINT which replaces the call to PRINT, and which will handle printing of the extra 'que', in tenses of the subjunctive.

Clearly, for composite tenses, the program will need to know how to conjugate the verb 'avoir'. Add a further supplementary function, PR-COMPOSE, which deals with printing of composite tenses.

2. Write a short program which takes as argument a list representing a sentence and which returns the sentence in the plural. Example:

```
(PL' (JE FINIS AVEC CE CHAPITRE))
            → (NOUS FINISSONS AVEC CE CHAPITRE)
```

or

```
(PL '(IL ME PARLE)) → (ILS ME PARLENT)
```

or again

```
(PL '(TU NE FOULERAS PLUS CETTE TERRE))
                    → (VOUS NE FOULEREZ PLUS CETTE TERRE)
```

3. An exercise for experts in algorithms. Write a program resolving the Hanoi Tower problem, but where discs are stacked on each of the needles at the start. The rules remain the same, with the exception that it is now allowed to put a disc on another *of the same size*. At the end, the target needle must contain all the discs in decreasing size order. Here is a diagram showing the initial and final states:

   Initial state:

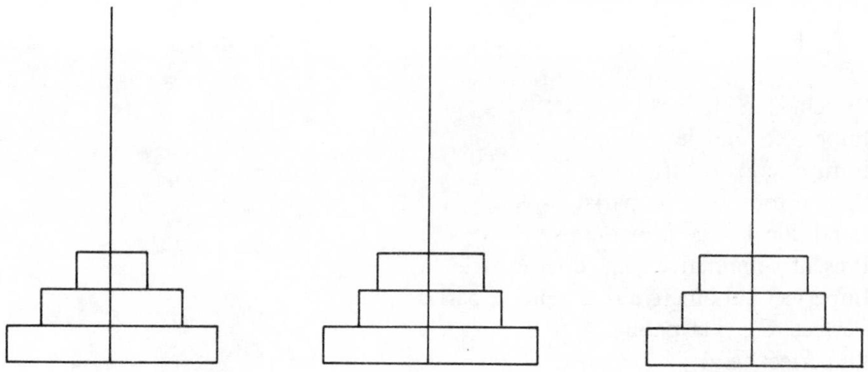

   Final state:

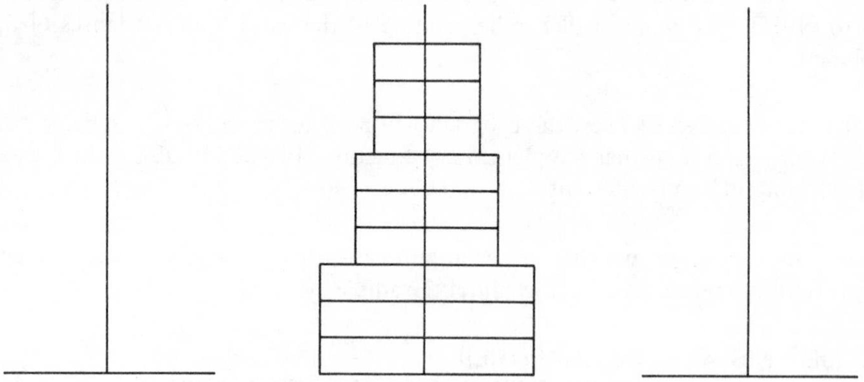

11 SOME OBSERVATIONS ON BINDING MECHANISMS

11.1 THE VARIOUS TYPES OF Le_LISP USER FUNCTIONS: DE AND DF[1]

11.1.1 The function DE

To define a user function, we use a standard LISP function DE. As we have seen, DE defines a function with a certain *name* and a certain number of *parameters*: as many parameters as there are variables in the list of variables.

When a call is made to such a function, each argument is evaluated and bound to the corresponding variable. For example, if we define function FOO as:

```
(DE FUN FOO (X Y Z) (CADR (PRINT (LIST X Y Z))))
```

we describe a function with name FOO, with three arguments called X, Y and Z respectively, which prints the values of this argument. In other words, if we call this function with:

```
(FOO (CAR '(A B C)) 1 2 )
```

the first thing that happens is that variable X is bound to atom A (the value of the first argument), variable Y is bound to the number 1 (the value of the second argument) and the third variable is bound to the value of the third argument, the number 2. Next, the body of the function is evaluated with the variables bound to these values. The effect is therefore to print the list:

```
(A 1 2)
```

and the result of this call will be the numeric value 1.

[1] This chapter may be omitted for readers who only want to have an introduction to Common-LISP.

The function DE therefore allows the definition of user functions with a fixed number of arguments, each of which is evaluated during a call. We may often need to define a function with an arbitrary number of arguments, which can take n arguments (with $n \in [1,\infty]$). Assume then that the standard function + will only take two arguments and that we want to have a function PPLUS which will add any number of integers. A first solution to this problem would be to define the function below:

```
(DE PPLUS (L)
    (IF (NULL L) 0
        (+ (CAR L)(PPLUS (CDR L)))))
```

This function requires a list of arguments when called. It therefore has to be called by:

```
(PPLUS '(1 2 3 4 5 6))
```

or

```
(PPLUS '(−1 2 −3 4 −5 6 −7 8))
```

In reality, it would be better to write the arguments one after the other, as below:

```
(PLUS 1 2 3 4 5 6)
```

or

```
(PLUS −1 2 −3 4 −5 6 −7 8)
```

In Le_LISP, we have only to write:

```
(DE PLUS L (PPLUS L))
```

with function PPLUS defined as above.

Examine the difference carefully between this function definition and those we have already met: instead of a *list of variables*, all we have is an *atom variable*.

If you define a function whose function-name is followed by an atom, this atom will, on a call being issued to the function, be linked to the *list of all the values of the different arguments*.

In function PLUS above, when the following call is made:

```
(PLUS 1 2 3 4 5 6)
```

the single variable L will be bound to the list of values of the various arguments, i.e. to the list:

(1 2 3 4 5 6)

This list is then sent to function PPLUS which expects a list of numbers and calculates their sum.

And here is another example: you have probably noticed that it is difficult to build up a new list starting from a series of elements. Each time we were in this situation, we had to write something like

(CONS *element*$_1$ (CONS *element*$_2$. . . (CONS *element*$_n$ ()). . .))

For example, in order to construct the list (A B C) from atoms A, B and C, we had to write:

(CONS 'A (CONS 'B (CONS 'C NIL)))

As we now know how to define functions with a random number of arguments, we can very simply write a function LIST which constructs a list made up of the values of its different arguments. Here it is:

(DE LIST L L)

It seems too simple to be true. Let us see what happens in the call:

(LIST 'A 'B 'C)

1. First variable L is bound to the list of values of the argument. The arguments are 'A 'B and 'C respectively. The values of these arguments are A, B and C, and variable L is therefore linked to the list (A B C).

2. The body of the function is simply the expression L, and therefore an evaluation of variable L. The value of this evaluation is the value of the call to function LIST. The value of the evaluation of a variable is, of course, the value to which it is bound. This call therefore returns as value the list (A B C), which is the list we wanted to construct.

Can you see the value of this binding mechanism?

Note, however, that there is a problem with this kind of function definition: there is no way we can carry out recursive calls to it. To see why, let us return to the small function PLUS above and see if we can avoid using the auxiliary function PPLUS:

```
(DE PLUS L
    (IF (NULL L) 0
        (+ (CAR L) (PLUS (CDR L)))))
```

This way of writing the function, though it seems it's only natural, is clearly wrong.

If you do not know why, do not read on straight away, but stop and try to find the error yourself.

In the function call:

(PLUS 1 2 3 4)

variable L is first associated with the *list* (1 2 3 4), by the simple binding mechanism used for this kind of function definition. Given that the list is not empty, LISP therefore proceeds with the evaluation of the line:

(+ (CAR L) (PLUS (CDR L)))

which should therefore calculate the sum of 1 and the result of the recursive call to PLUS with argument (2 3 4), the CDR of the value of variable L. But, horror of horrors, we are no longer submitting a series of numbers to function PLUS, but a list. We know that L is bound to the list of the values of the arguments: L will therefore receive a list consisting of just one element, the value of the argument (2 3 4), which is certainly not what we wanted to do (what is the value of (2 3 4)? certainly not the sequence of numbers 2, 3 and 4).

This is the difficulty which led us, when we wrote function PLUS, to leave the loop (the repetition) to an auxiliary function, in this case function PPLUS.

You should remember that it is impossible to make simple recursive calls each time you build up a function with an arbitrary number of arguments. Later in this book we shall, of course, study a solution to the problem of making recursive calls to this type of function.

In Le_LISP, user functions defined with function DE are called EXPRs, and in old dialects of LISP, such as MACLISP, functions with any number of arguments are called LEXPRs or NEXPRs.

11.1.2. The function DF

In the same way as it is useful to have a function with any number of arguments which evaluate their arguments, we sometimes need functions with any number of arguments which do *not* evaluate their arguments. These can be defined using the user definition function DF, defined as follows:

(DF *function-name (variable) body-of-function*)

Syntactically, DF is identical to DE, DF may have several variables. The binding is made *exactly* in the same way as for DE, except that the various arguments are *not* evaluated.

Here is the simplest example:

```
(DF QLIST L L)
```

This function is very similar to function LIST above, since, as with LIST, the body of the function is simply the evaluation of the single parameter variable. Given that the arguments will *not* be evaluated, the call

```
(QLIST A B C D)
```

therefore returns the list of arguments as they stand, i.e.:

```
(A B C D)
```

Note that the various arguments *have not been* quoted. If the call is made as follows:

```
(QLIST 'A 'B 'C)
```

the result is the list:

```
('A 'B 'C)
```

In what follows, we shall see how useful such functions can be, and how one or other of the arguments may be evaluated.

The user functions defined with function DF are called FEXPRs.

11.1.3 Exercises

1. Define a function TIMES, with an arbitrary number of arguments, which calculates the product of all these arguments. Examples:

```
(TIMES 1 2 3 4)                                        → 24
(TIMES (CAR '(1 2 3))(CADR '(1 2 3))(CADDR '(1 2 3)))  → 6
```

2. Define a function SLIST, with an arbitrary number of arguments, which builds a list containing *the values* of the arguments of even rank. Example:

```
(SLIST 1 2 3 4 5 6)       →(2 4 6)
(SLIST 1 2 3 4 5 6 7)     →(2 4 6)
(SLIST 1 'A 2 'B 3 'C)    →(A B C)
```

3. Define another function QSLIST, with an arbitrary number of arguments, which

constructs a list containing the arguments of even rank *as they stand*. Examples:

```
(QSLIST A B C D E F)           →(B D F)
(QSLIST A B C D E F G)         →(B D F)
```

4. Finally, define a function MAX which finds the greatest in a series of numbers and a function MIN which finds the smallest in a series of numbers. Here are a few examples of calls:

```
(MAX 1 2 10 -56 20 6)                              → 20
(MIN 1 2 10 -56 20 6)                              → -56
(MAX (MIN 0 1 -34 23 6)(MIN 36 37 -38 476 63))     → -34
```

11.2 DETAILS ABOUT COMMON-LISP FUNCTION DEFINITIONS

In Le_LISP it is the structure of the variable list which determines the bindings of the variables.

To see how this functions, let us go back to the PLUS function defined at the beginning of this chapter. Remember that our problem was to define a function with a variable number of arguments. Supposing that the function + takes only two arguments, we wanted to write a function PLUS computing the sum of an arbitrary amount of numbers. Given that each parameter of the variable list is normally bound to one and only one argument, we needed a construction simulating an infinite number of parameters. In Le_LISP, the solution to this problem is to bind a variable (*the unique* variable representing the 'list' of variables) to the list of values of *all* the arguments.

In Common-LISP, the same effect may be obtained through the use of special *keywords* inside the variable list. So, in Common-LISP, this same function PLUS would be defined as follows:

```
(DEFUN PLUS (&REST L) (PPLUS L))

(DEFUN PPLUS (L)
        (IF (NULL L) 0 (+ (CAR L)(PPLUS (CDR L)))))
```

The special symbol &REST, is a *keyword* telling Common-LISP to bind to the variable following this keyword the list of values of *all* the remaining parameters. Thus, if we call the function PLUS like:

```
(PLUS 1 2 3 4 5)
```

the variable L – the variable immediately following the keyword &REST – will be bound to the list

```
(1 2 3 4 5)
```

and this list will be sent to the function PPLUS which incrementally computes the sum of these numbers, i.e. 15.

Actually, no real Common-LISP programmer would define the function PLUS in the way we had done, that is with the auxiliary function PPLUS: this was just a way to make the transition from Le_LISP to Common-LISP. In Common-LISP we would use the function REDUCE and write PLUS as follows:

```
(DE PLUS (&REST L) (REDUCE #'+ L))
```

which does exactly the same computation, but is more efficient and more elegant – where elegant means just: it is shorter and takes less list space.

Note that we quoted the + function in a very peculiar way: instead of just using the quote character (') we preceded the quote character by a sharp sign (#). This is not necessary, but it's a means to tell Common-LISP that the quoted object is referring to a function, so Common-LISP can handle this more efficiently than with the single quote character.

The rest parameter may be combined with normal parameters if it is the last parameter in the variable list. For example, if we want a function to multiply the sum of an arbitrary amount of numbers by a given value, we could write:

```
(DEFUN MULT-PLUS (FACTOR &REST L)
       (* FACTOR (REDUCE #'+ L)))
```

which, when called with:

```
(MULT-PLUS 10 1 2 3 4 5)
```

will bind the variable FACTOR to the value 10, the value of the first argument, and the variable L to the list (1 2 3 4 5), the list of the values of the rest of the arguments. Naturally, this function call will return the value 150.

As shown in the preceding paragraph about Le_LISP's binding mechanisms, similar remarks hold for Common-LISP, especially the remark that it is not possible to call recursively a function with a rest parameter. Consider for example the following definition of the PLUS function:

```
(DEFUN PLUS (&REST L) (COND
       ((NULL L) 0)
       (T (+(CAR L) PLUS (CDR L))))))
```

This function is erronous, for exactly the same reasons as the corresponding function in Le_LISP was erronous: since the parameter L is bound to the *list* of values of all the parameters, in the recursive call to PLUS, the variable L will be bound to the list containing the list resulting from the evaluation of (CDR L). This is surely not what was intended, and it is surely not an arbitrary number of arguments: there is just one!

Sometimes one doesn't need an arbitrary number of arguments, but a few, a well defined number of optional arguments. For example, remember our definition of the function REVERSE given in Chapter 6. Here it is again:

```
(DEFUN REVERSE (L RES)
   (IF (NULL L) RES
      (REVERSE (CDR L) (CONS (CAR L) RES))))
```

This function has always to be called with two (2) arguments, which is conceptually a little awkward: if we want to reverse a list, it is *one* list we want to reverse, and it shouldn't be our obligation to know about the *how* of the implementation, that is we don't want to be obliged to give an initial value for the accumulator RES. Clearly, instead of

```
(REVERSE '(A B C) NIL)
```

we would like to be able to call it just like:

```
(REVERSE '(A B C))
```

But, with the definition above, this last call would produce an error: Common-LISP would correctly inform us of the fact that one argument is missing. With all what we know until now, the only way to circumvent this problem would be to rewrite the function in the following form:

```
(DEFUN REVERSE (L) (REV L NIL))

(DEFUN REV (L RES)
   (IF (NULL L) RES
      (REV (CDR L) (CONS (CAR L) RES))))
```

Here the first function, REVERSE, just serves to initialize the accumulator RES of the function REV, and it is REV which does all the work. This is definitely a solution, but not a very elegant one.

The problem we are handling here is that one of the arguments is *optional*: RES, in the original version of REVERSE, is for the first call superfluous, but for all the recursive calls it is absolutely necessary.

The Common-LISP keyword &OPTIONAL exists for just this reason: it indicates that the parameters following this keyword may or may not receive values through the arguments of a call to the function. For example, the REVERSE we would write using this keyword would be like:

```
(DEFUN REVERSE (L &OPTIONAL RES)
      (IF (NULL L) RES
         (REVERSE (CDR L) (CONS (CAR L) RES))))
```

If we call this function like:

```
(REVERSE '(A B C))
```

the variable L will be bound to the list (A B C), and the optional parameter RES will be bound to NIL, which is the default value of an optional parameter. Naturally, we still could call REVERSE with an explicit second argument (this is done actually in all recursive calls), so, for example, the call:

```
(REVERSE '(A B C) '(A B C))
```

would return the list (C B A A B C).

Let us take another example. Here is the function FOO which multiplies a number X by another number Y. If Y is not given, X is multiplied by 2:

```
(DEFUN FOO (X &OPTIONAL Y)
         (* X (IF Y Y 2)))
```

and here are two examples of its use:

```
(FOO 24)          → 48
(FOO 24 5)        → 120
```

FOO is a function *allowing* two arguments: the first one, X, is a *required* argument, if we don't supply it, Common-LISP will give an error-message. The second one Y is optional, if we don't supply it, Common-LISP will initialize the argument to NIL.

It is possible to explicitly determine default values for optional arguments. A default value is the value a variable will take if no arguments are supplied. In our FOO function, we determined the default value 2 inside the body of the function, through the evaluation of the expression (IF Y Y 2). We could have written the same function more elegantly as follows:

```
(DEFUN FOO (X &OPTIONAL (Y 2)) (* X Y))
```

Here we use a Common-LISP feature permitting us to follow the keyword &OPTIONAL not only by the name of a parameter, but by a list composed of the name of the parameter and its default value. This is very useful every time we need a default value different from NIL.

Naturally, one can combine, in a parameter list, required, optional and rest parameters – *always in this order*: first the required parameters, then the optional parameters and at the end of the variable list the rest parameter. To see how the binding is done here is a simple example:

```
(DEFUN BAR (X &OPTIONAL Y &REST Z) (LIST X Y Z))
```

In this function, the required parameter is X, the optional parameter is Y and the rest parameter is Z. Following are a few examples of the function BAR:

```
(BAR 1 2 3)             →(1 2 (3))
(BAR 1 2)               →(1 2 NIL)
(BAR 1)                 →(1 NIL NIL)
(BAR 1 2 3 4 5 6)       →(1 2 (3 4 5 6))
```

Common-LISP permits a third argument in the list following the &OPTIONAL keyword. This third argument permits us to distinguish, inside the body of the function, if the optional parameter was supplied or not. For example, in the definition of FOO below:[2]

```
(DEFUN FOO (X &OPTIONAL (Y 2 YFLAG))
        (IF (NULL YFLAG) (PRINC "the default factor is 2"))
        (* X Y))
```

YFLAG will be bound to NIL when no second argument has been supplied, otherwise it will be bound to T. Here are some examples:

```
?  (FOO 1)
   the default factor is 2
=  2

?  (FOO 2 3)
=  6

?  (FOO 100)
   the default factor is 2
=  200
```

With the two keywords available, OPTIONAL and REST, we can organize our variable lists in rather flexible ways. There is just one remaining problem: what can be done if one has more than one independent optional parameter?

Let us take an abstract example:

```
(DEFUN EXAMPLE (&OPTIONAL X Y) . . . )
```

[2] This example is inspired by a similar example in Wilensky's book, *Common LISP Craft*, 1984.

If we call this function, with

```
(EXAMPLE)
```

X and Y will be bound to NIL since neither an argument for X nor one for Y is supplied. The call:

```
(EXAMPLE 1)
```

will bind X to 1 and Y to the default value NIL. The call:

```
(EXAMPLE 1 2)
```

will bind X to 1 and Y to NIL. There is no way to only supply an argument for Y. If we want to do that, we are obliged to give an explicit first argument of NIL, like in:

```
(EXAMPLE NIL 2)
```

even if it is only the binding of Y which is of interest. Though this looks somewhat esoteric here, it happens rather often during the construction of LISP programs that one has independent optional arguments. In all these cases to give all the previous arguments with an explicit NIL, supposing that the user knows the order of all the arguments, is really too awkward. That is why Common-LISP has a third keyword: the keyword &KEY. To refer to a keyword parameter while calling the function, the user supplies as an argument the *name of the keyword (preceded by the ':' character)* and following this the argument for this keyword parameter.

In the example above, this could be written, in the definition, as:

```
(DEFUN EXAMPLES (&KEY X Y) . . . )
```

and a possible call would look like:

```
(EXAMPLE :X 1 :Y 2)
(EXAMPLE :Y 2 :X 1)
(EXAMPLE :Y 2)
(EXAMPLE :X 1)
(EXAMPLE)
```

The first two examples above bind X to 1 and Y to 2. The third example call binds X to NIL (since it is not supplied as an argument) and Y to 2. The fourth example call binds X to 1 and Y is defaulted to NIL. In the last example call, since neither of the arguments is supplied, both X and Y, are initialized to NIL.

Like all other optional parameters, &KEY optional parameters may be given with an initial value different from NIL and a flag indicating if an argument has been supplied or not. The syntax is exactly the same as for the keyword &OPTIONAL.

Just construct some functions using these different keywords, and try to understand how the binding is functioning.

To end this chapter, below are two function definitions with some example calls:

```
?   (DEFUN EXAMPLE1 (&KEY X Y) (LIST X Y))
=   EXAMPLE1

?   (EXAMPLE1 :X 1 :Y 2)
=   (1 2)

?   (EXAMPLE1 :Y 2 :X 1)
=   (1 2)

?   (EXAMPLE1 :Y 1)
=   (NIL 1)

?   (EXAMPLE1)
=   (NIL NIL)

?   (DEFUN EXAMPLE2 (A &OPTIONAL (B 3)
                         &REST X
                         &KEY C (D A))
               (LIST A B C D X))
=   EXAMPLE2

?   (EXAMPLE2 1)
=   (1 3 NIL 1 NIL)

?   (EXAMPLE2 1 2)
=   (1 2 NIL 1 NIL)

?   (EXAMPLE2 :C 8)
=   (:C 8 NIL :C NIL)

?   (EXAMPLE2 1 2 :C 8)
=   (1 2 8 1 (:C 8))

?   (EXAMPLE2 1 2 :D 3)
=   (1 2 NIL 3 (:D 3))

?   (EXAMPLE2 1 2 :D 8 :C 9 :D 10)
=   (1 2 9 8 (:D 8 :C 9 :D 10))
```

12 FUNCTIONS EVAL AND APPLY

In LISP, lists are both data that has to be processed and the programs themselves. There is no reason, therefore, why programs should not be handled like perfectly ordinary lists. For example, the list:

```
(CDR '(HEAD HEAL TEAL TELL TALL TAIL))
```
[1]

could either be *evaluated*, which would have the effect of returning the value

```
(HEAD TEAL TELL TALL TAIL)
```

or be the argument for another function, as in:

```
(CDR (CDR '(HEAD HEAL TEAL TELL TALL TAIL)))
```

then again, it can be processed like an ordinary list, as is the case, for example, in:

```
(CONS 'CDR (CONS '(CDR '(HEAD HEAL TEAL TELL TALL TAIL)) ()))
```

which is itself a function call. In LISP, no distinction is made between the representation of data and the representation of programs. It is therefore possible to use a program to construct other programs automatically.

The problem that is posed next is to find a way of *running* the programs constructed in this way, i.e. how to go from the LISP expression:

```
(CDR (CDR '(HEAD HEAL TEAL TELL TALL TAIL)))
```

to its value:

```
(TEAL TELL TALL TAIL)
```

The two LISP functions allowing us to calculate LISP values from expressions constructed in LISP, are called EVAL and APPLY. It is these two functions that we shall study in this chapter.

---

[1] This is one of the solutions to Lewis Carroll's problem of finding ways to go from head to tail.

12.1 FUNCTION EVAL

The EVAL function takes one argument and returns the value of the evaluation of the value of that argument. This may sound complicated. The best way to explain it is to look at some examples. The value of the expression

```
(CONS 'CDR (CONS '(CDR '(HEAD HEAL TEAL TELL TALL TAIL))()))
```

is the list

```
(CDR (CDR '(HEAD HEAL TEAL TELL TALL TAIL)))
```

If we now consider this list as an expression to evaluate, its value, as we said above, is

```
(TEAL TELL TALL TAIL)
```

which is also the value of the expression:

```
(EVAL (CONS 'CDR (CONS '(CDR '(HEAD HEAL TEAL TELL TALL TAIL)) ())))
```

This means that function EVAL allows us to *force* a further evaluation. Let us take another example. If we submit the following expression to LISP

```
'(+ (1+ 4)(* 4 4))
```

LISP will reply with a value of the expression which, as the whole expression is *quoted*, is the expression itself, i.e.:

```
(1+ (1+ 4)(* 4 4))
```

If, on the other hand, we submit the following expression to LISP

```
(EVAL '(+ (1+ 4)(* 4 4)))
```

LISP will return the value 21, i.e. the value of the evaluation of this expression. EVAL is therefore the opposite of function QUOTE: while QUOTE prevents the evaluation of its argument, EVAL forces such an evaluation.

Here then is the definition of EVAL:

(EVAL *expression*) → *value (expression)*[2]
EVAL evaluates the value of its argument.

[2] Notation *value* (*x*) reads as: calculate the value of the value of *x*.

Take another example. Say we want to write a simple calculator. This means we want to be able to submit expressions of the following form to LISP:

```
(1 + 2 + 3)
(1 + (2 * 3))
((1 + 2) * 3)
(1 * 2 * 3)
    . . .
```

Of course, the program must reply with the value of these expressions. One way of constructing the calculator would be to begin by translating the expressions into a form understandable to LISP, and then evaluate them. In other words, the arithmetic expressions above would have to be translated into the following form:

```
(+ 1 (+ 2 3))
(+ 1 (* 2 3))
(* (+1 2) 3)
(* 1 (* 2 3))
    . . .
```

Here is a function which translates arithmetic expressions written in *infix* notation into arithmetic expressions written in *polish prefix* notation (without, however, taking account of standard operator priorities), and therefore into forms that LISP can evaluate:

```
(DEFUN PREFIX (L)
       (IF (ATOM L) L
         (CONS (CADR L)
              (CONS (PREFIX (CAR L))
                   (IF (MEMQ (CADDR (CDR L ))'(+ * - /))
                     (CONS (PREFIX (CONS (CADDR L))
                               (CONS (CADDR (CDR L))
                                   (CDDR (CDDR L))))) ())
                               (CONS (PREFIX (CADDR L))()))))))))
```

Here are two examples of calls to this function:

```
(PREFIX '(1+2+3))                → (+ 1 (+ 2 3))
(PREFIX '(1+2+(3*4)+(80/10)))→ (+ 1 (+ 2 (+(* 3 4)(/ 80 10))))
```

These are *forms*, and therefore calls to LISP functions. In order to return a result, all we need to do is to *evaluate* these forms, which simply requires a call to function EVAL. Here then is the calculator function:

```
(DEFUN CALCULATE (L) (EVAL (PREFIX L)))
```

If we call this new function for the examples above, we obtain:

```
(CALCULATE '(1+2+3))                       → 6
(CALCULATE '(1+2+(3 * 4)+(80 / 10)))       → 23
```

and finally

```
(CALCULATE '(2 * 3 * (1 - 5) + 10))        → 36
```

Note that the call (EVAL *x*) generates *two* evaluations of *x*:

1. First *x* is evaluated because this expression is the argument of a function.

2. Next the value calculated is evaluated again because of the explicit call to EVAL.

To make this clearer, look at the following expression:

```
(LET ((LE 'THE) (X 'LE))(PRINT X (EVAL X)))
```

which prints

```
LE THE
```

and returns the atom THE as value.

First, variable LE is bound to the atom THE and the variable X is bound to atom LE. The value of expression EVAL X is THE, which is the value of variable LE, itself the value of variable X.

12.2 FUNCTION APPLY

The other explicit evaluation function available in LISP is APPLY. Its name is derived from the fact that a function is applied to a series of arguments.

APPLY is defined syntactically as follows:

(APPLY *function list*)

It has two arguments which are both evaluated. The first argument must be an expression which returns a function and the second must produce a list whose elements will be the arguments for the function passed as first argument. APPLY then applies the function to the list of arguments.

Here are a few examples:

```
(APPLY '+ '(20 80))                                          → 100
(APPLY 'CAR '((DO RE MI)))                                   → DO
(APPLY (CAR '(CONS CAR DCR)) (CONS '(DO RE) '((MI FA)))) → ((DO RE) MI FA)
```
[3]

It all looks *as though* the function APPLY does two things:

1. First, it builds up a function call using its two arguments. For example:

   (APPLY '+ '(20 80)) is converted into (+ 20 80)

   and

   (APPLY 'CAR '((DO RE MI))) is converted into (CAR '(DO RE MI))

2. Next, the expression built up in this way is evaluated. You should remember that:

1. The function evaluates both its arguments;

2. It expects a *list* of arguments: there is, therefore, always an extra pair of parentheses around the arguments;

3. After having evaluated the two arguments it *applies* the function, returned by evaluating its first argument, to the arguments in the list, which is itself returned by evaluating the second argument.

The relation between EVAL and APPLY can be expressed functionally, by the equation

(APPLY *function list*) = (EVAL (CONS *function list1*))

where *list1* is the value of *list* (the second argument to APPLY) with each element quoted.

The main value of APPLY is that it resolves the problem that we had in the previous chapter, i.e. how to call a user function of the NEXPR type recursively.

Remember that the function that we gave as an example was:

---

[3] Remember that in Common-LISP it is preferable to quote functional objects with the sharp sign (#). So, for example,

```
(APPLY '+ '(20 80))
```

is better written as

```
(APPLY #'+ '(20 80))
```

```
(DEFUN PPLUS (L)
        (IF (NULL L) 0
           (+ (CAR L)(PPLUS (CDR L)))))
```

(DE PLUS L (PPLUS L)) in Le_LISP

or

(DEFUN PLUS (&REST L)(PPLUS L)) *in Common-LISP*

We needed the auxiliary function PPLUS because of the binding between variable L and the *list* of the arguments. Thanks to function APPLY, which provides the argument of the function passed as first argument in a list, we can now rewrite this function in the following way:

```
(DEFUN PLUS (&REST L)
   (IF (NULL L) 0
      (+ (CAR L) (APPLY 'PLUS (CDR L)))))
```

We could, of course, also have written PLUS as follows:

```
(DEFUN PLUS (&REST L)
        (IF (NULL L) 0
           (+ (CAR L)(EVAL (CONS 'PLUS (CDR L))))))
```

But this second form requires an additional call to function CONS, and is therefore likely to be more costly in execution time. Note, however, that, in Le_LISP, if we want to make recursive calls with functions of the FEXPR type (i.e. user functions defined by means of DF), we have to write the call in this form, since APPLY *will not accept* such functions as functional arguments.

The first argument for APPLY can be any LISP expression. All that is needed is that this expression should return a function (with the restriction that we have just mentioned).
Consequently, if P is defined as:

```
(DEFUN P (X Y) (+ X Y))
```

all the forms below are correct and will calculate the sum of 1 and 2:

```
(APPLY '+ '(1 2))
```

```
(APPLY (CAR '(+ - * /)) '(1 2))
(LET ((X '+))(APPLY X '(1 2)))
(APPLY 'P '(1 2))
```

12.3 EXERCISES

1. Repeat each of the three exercises in the previous chapter and alter the solutions so as to include recursive calls using function EVAL or function APPLY.

2. Write a small translation program from French into English. For the moment, ignore any grammar aspect and concentrate on word by word translation.

   Do this by writing a function which takes each word you want to translate and returns the translated word. Here, for example, is function CHAT, which translates the French word "chat" into its english equivalent "cat":

   ```
   (DEFUN CHAT () 'CAT)
   ```

   The translation function will then have to use these word functions, to carry out small translations such as:

   le chat mange → the cat eats
   le chat mange le souris → the cat eats the mouse
   le souris a vole le fromage → the mouse has stolen the cheese

   Clearly, we must absolutely avoid defining translation functions for words which correspond to the names of standard functions, such as, for example, DEFUN.

13 INPUT/OUTPUT (PART TWO)

It is indispensable to have functions allowing us to read the objects on which we want to work. As long as all working objects are either inside a program, or given as arguments in function calls, the LISP system deals with data reading implicitly. On the other hand, if we want to introduce additional data as the program is executed, we have to use explicit calls to read functions. Any LISP system has at least one such function: function READ. It has no arguments, and returns as value the LISP expression read from the input stream.

Before we go any further with the READ function, let us take a look at the problems associated with input/output. In normal interaction with your LISP system, you enter functions and data on the keyboard of your terminal and receive results printed on the terminal screen. This means that in the standard case, the *entry file* for LISP is the keyboard, and the *output file* for LISP is the screen. Things do not have to be this way. You may want to write the results of calculations to a disk file, to keep them for later use, or you may want to write them to a magnetic tape, a cassette or even to punched tape (if such tape still exists). In the same way, your input may not necessarily come from the keyboard alone, but again from a disk file, from a card reader, from a punched tape reader, from a magnetic tape reader or from a cassette. In these cases, you have to tell LISP beforehand that you want to change the *input file* or the *output file* (we shall see later how this is done). Remember, for the moment, that input/output are *associated* with *input/output files* which can be determined by program, and which by default (and until further notice) are the keyboard, for data entry, and the display screen for outputting results.

13.1 THE GENERAL INPUT FUNCTION: READ

Let us now return to the READ function. Here is its definition:

(READ) → *expression read*

READ is a function without an argument which reads a LISP expression (a list or an atom) from an input stream and which returns that expression as value.

The READ function does not therefore evaluate the expression it reads.

This is a function of fundamental importance within the LISP interpreter itself. An interpreter is a program which allows the machine to understand a programming language. A LISP interpreter is a program that allows the computer to understand the LISP language. It is made up of three parts (at least): a part which reads what the user (the programmer) enters into the machine, a part which computes the value of what it has read, and a third part which displays the results of its computations. Normally, an interpreter repeats these three operations infinitely:

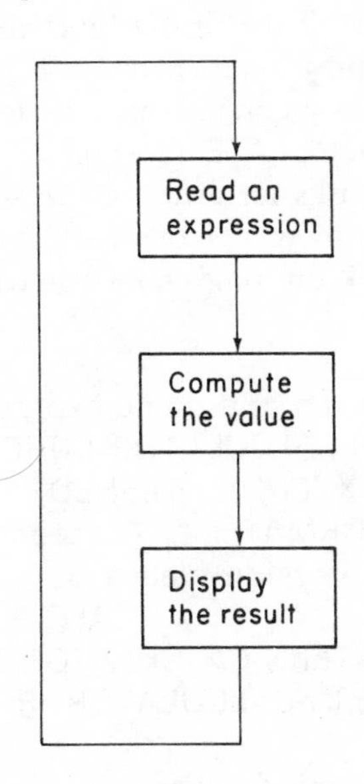

Here is a little LISP interpreter written in LISP:

```
(DEFUN LISP ()
    (FORMAT T "→ ~A~%" (EVAL (READ)))
    (LISP))
```

This is a program with no end. Here is a short trace of the use of this function:

?(LISP)	*the question mark is printed by LISP, indicating that it is expecting an expression to evaluate. This launches function LISP defined above.*
? 1	*again it is LISP that displays the question mark. More precisely: the function* READ *prints a ? each time it is expecting entry from the terminal. Here, it has been given the numeric expression 1 to evaluate.*
→ 1	*LISP function has evaluated an expression and printed its value, preceded by the '→' sign.*
?(CAR '(A B C))	*LISP is calling itself recursively and is therefore asking for another expression to evaluate. We ask for the* CAR *of the list* (A B C).
→ A	*again, LISP gives us the result of the evaluation (atom* A*), preceded by the sign '→'*

?	*and so on. Clearly, what is happening is exactly the same way as in an interaction with the ordinary LISP interpreter, except that the result is preceded by the '→' sign and not by the '=' sign.*

To give another example of the use of the READ function, let us return to the little calculator we described in the preceding chapter. The main function is called CALCULATE. We shall construct a program which reads one arithmetic expression after another, sends it to the CALCULATE function to calculate the value of the expression, prints that value, and asks for a new expression. This goes on until we enter the atom END, which takes us out of the calculator and back to the normal state. This is another interpreter, but this time an interpreter of arithmetic expressions:

```
(DEFUN CALCULATOR ()
   (FORMAT T "enter an arithmetic expression please ")
   (LABELS ((INTERN-CALCULATOR (X) (COND
                  ((EQ X 'END) 'FINISHED)
                  (T (FORMAT T "~A~% another arithmetic
                      expression please "
                                 (CALCULATE X))
                      (INTERN-CALCULATOR (READ))))))
                  (INTERN-CALCULATOR (READ)))))))))
```

Below we give a trace of the use of this function:

```
?(CALCULATOR)                                the function is launched
enter an arithmetic expression please        the prompt
?(1 + 2 + 3)                                 the first arithmetic expression
6                                            the result
another arithmetic expression please         a further prompt
? (1 + 2 + (3 * 4) + (80 / 10))
23
another arithmetic expression please
?(1 + 2 + 3 * 4 + 80 / 10)
39
another arithmetic expression please
? END                                        entry indicating that we want to end the
                                             session on the calculator
= FINISHED                                   exit returning atom FINISHED as value
?                                            here we are back at LISP
```

13.2 OTHER INPUT/OUTPUT FUNCTIONS

Let us now turn back to read functions. Input takes place, like output, in two stages: first a buffer is filled (the input buffer) with information from an input peripheral;

then the buffer is analysed by LISP. Analysis does not normally start until the complete line has been read.

Function READ therefore implements two operations: if the input buffer is empty, it launches the auxiliary program which handles physical READ operations (physically reading the next record in an input stream), and then it reads a LISP expression from the buffer (the LISP expression may be an atom or a list). If the expression is not complete at the end of the record (a record is a line of text if input is taking place from the terminal), READ launches a further physical read operation.

Imagine that you are executing a call to function READ and that you then type the following line:

```
(CAR '(A B C))(CDR '(A B C)) <cr>
```

where <cr> means a carriage return.

This line will be placed in the input buffer by the physical READ program. When analysis begins of the line that has been read, the buffer has the following contents:

```
| (CAR '(A B C)) (CDR '(A B C)) <cr>                              |
 ↑
```

with a pointer to the start of the buffer. This pointer is represented here by the arrow below the rectangle representing the buffer.

After execution of the READ, only the *first* expression will have been read. The buffer pointer will be positioned at the end of the first expression, at the beginning of the sequence of characters which has not yet been analysed.

```
|(CAR '(A B C)) (CDR '(A B C)) <cr>                               |
               ↑
```

It will need a further call to the READ function to *read* the second expression too. (Can you see the difference between physical reading and LISP reading?) Clearly, if LISP meets a carriage return (shown here by the sign <cr>) while reading is taking place, it relaunches the auxiliary physical read function.

The important thing is to have a clear idea of the difference between physical reading and analysis of the text that has been read. Function READ triggers *both* processes.

If, while the buffer contains

```
|(CAR '(A B C)) (CDR '(A B C)) <cr>                               |
               ↑
```

we call the function READ-CHAR[1], Common-LISP will read the next *character* in the input buffer, *without carrying out any analysis on it*. Here, 'read' means returning the next character as value and advancing the position of the input buffer pointer one place. The buffer would then be in the following state:

```
(CAR '(A B C)) (CDR '(A B C)) <cr>
               ↑
```

The character that has been read is (in this case) the character <space>, normally a separator character that is ignored.

Here is the definition of the function READ-CHAR.

(READ-CHAR) → *character*
: reads the next character in the input stream and returns it as value in the form of a single character atom.
: READ-CHAR reads *any* character.

This function is very useful if you want to read characters which are not accessible by means of the READ function, such as the '(', ')', ';', '.', <return> or <space>.

The function PEEK-CHAR[2] is exactly like the function READ-CHAR but it does not advance the input buffer pointer. The character read therefore remains available to the next READ-CHAR, PEEK-CHAR or READ. Here is its definition:

(PEEK-CHAR) → *character*
: reads the next character in the input stream and returns it as value in the form of a single character atom.
: PEEK-CHAR reads *any* character and leaves it available in the input stream for further reads.

In addition to the functions READ-CHAR and PEEK-CHAR, Common-LISP offers the function READ-CHAR-NO-HANG. This function is like READ-CHAR *if* a character is available, otherwise it returns NIL. This is very useful if one wants to construct a program which reacts immediately, for example: as soon as one has typed a character on the keyboard.

The output function corresponding to READ-CHAR is WRITE-CHAR, which takes a character as an argument and prints it on the out-put stream.

---

[1] The function READ-CHAR of Common-LISP is called READCH in Le-LISP.

[2] The function PEEK-CHAR of Common-LISP is called PEEKCH in Le-LISP.

Le_LISP makes two immediate input/output functions available to you: TYI and TYO. *Immediate* means that the functions do not use input/output buffers. Here are their definitions:

(TYI) → *numeric value*
this function reads a character from the *terminal* and returns the ascii value of the character *immediately* after the character is typed. There is no need to end the line with a <return> The character type does not appear on the screen: we say that *character echoing has been inhibited.*

(TYO *n*) → *n*
This function displays the character corresponding to code *n* in ASCII, at the current cursor position on screen.

Note that the function TYO will take any number of arguments and displays all the corresponding characters.

These two functions are complementary, so if you want an echo (i.e. if you want the character typed to be displayed) for a call to function TYI, you have only to write

```
(TYO (TYI))
```

TYO allows you to output any character on your terminals. For example, if you want to produce a little beep on your terminal, you simply have to call (TYO 7), since 7 is the ASCII code for the <bell> character.

Naturally, if you want to do the same (i.e. produce a beep on the terminal) in Common-LISP, you have to use WRITE-CHAR with <ctrl>G as argument. Such that

```
(WRITE-CHAR #\^G)
```

where the ^G is produced in holding down the <ctrl> key while typing the character G, will produce a little beep on your terminal.

Clearly, these last four functions are the most elementary input/output functions in LISP. Any interface possible can be written with them.

After all this theory, we must absolutely return for a while to programming practice, in order to see how these functions can actually be used.

The Le_LISP Function TYI is very often used for programming interactive games (because its effect is immediate, once the character requested has been typed, so programs using this function are very sensitive). It is also used to prevent displayed information scrolling too fast (since the machine may *wait* for a character to be typed, but which character you enter is unimportant).

To illustrate the latter point, here is function COUNT, which prints all numbers fro
X to 0, one below another:

```
(DEFUN COUNT (X)
        (IF (LE X 0) "BANG"
           (PRINT X)
               (COUNT (1 - X))))
```

The call (COUNT 4) will give:

```
4
3
2
1
= BANG
```

The problem with this function is that a call such as (COUNT 1000) will produce print of the successive numbers that scrolls by too quickly to be read off the screen. To prevent this, in the second version a little counter has been included which count the lines printed. After printing 23 lines[3] the program waits for you to type som character before it continues. In this way you can interactively determine the speed c the display:

```
(DE COUNT (X)
   (LETN INTERNAL-COUNT ((X X) (Y 1))
          (IF (LE X 0) "BANG"
             (PRINT X)
             (INTERNAL-COUNT (1- X)
                         (IF (< Y 23) (1+ Y) (TYI) 1)))))
```

The differences between this Le_LISP version of the counting function and th Common-LISP version just following, is that TYI does not echo, i.e. print the characte typed, and that TYI does not look in a buffer, so it is not necessary to finish typin with a carriage return which is necessary for READ-CHAR.

```
(DEFUN COUNT (X)
   (LABELS ((INTERN-COUNT (X Y)(COND
               ((<= X 0) (TERPRI) "bang")
               (T (PRINT X)
                  (INTERN-COUNT (1- X) (COND
                                  ((< Y 23) (1+ Y))
                                  ((READ-CHAR) 1)))))))
         (INTERN-COUNT X 1)))
```

[3] Incidentally, number 23 is not only one less than the number of lines on most classical terminals, bu also the first number *n* such that the probability that two people out of *n* have the same birthday is greate than 50%.

Let us now construct a small LISP interpreter using a read function which keeps statistics on the use of atoms, while it reads. Each atom will be associated with an occurrence counter which is updated during reading.

In the first place, let us construct a new read function, STAT-READ

```
(DEFUN STAT-READ ()
(LET ((X (PEEK-CHAR))) (COND
     ((EQUAL X #\( ) (READ-CHAR) (READ-LIST))
     ((EQUAL X #\) ) (READ-CHAR) 'CLOSING-PARENTHESIS)
     ((MEMBER X '(#\Space #\Newline #\Return #\Tab))
                    (READ-CHAR) (STAT-READ))
     (T (READ-ATOM)))))
```

This function starts, after a call to function PEEK-CHAR with a tentative look at the input buffer: if the character waiting to be read is an opening parenthesis, it calls function READ-LIST, a function whose task it is to read lists. If the character at the head of the buffer is a closing parenthesis, the character is removed from the buffer (using a call to function READ-CHAR) and STAT-READ is exited, returning the message CLOSING PARENTHESIS. In cases where we find a standard separator at the head of the buffer, we eliminate it and continue to examine the remaining characters. If none of these cases occurs, we can be sure we are dealing with an atom, and function READ-ATOM takes over to deal with reading the atom.

Note that there are several special characters:

#\Space   #\Newline   #\Return   #\Tab

which are convenient ways of writing the characters <space> <new line>, <return> and <tabulation>. These are just *names* for some special characters. They exist mainly for keeping the programs using this characters readable.

Look carefully at the use of function PEEK-CHAR: it is this function which allows analysis of the various distinct cases, *without* actually affecting the character analysed. The character remains available for further reading. After all, the character will have to be analysed when we go on with reading the atom: it will be the first character in its Print-Name.

Function READ-LIST reads one element after another in the list, until it meets a closing parenthesis, and then actually constructs that list. Here it is:

```
(DEFUN READ-LIST ()
   (LABELS ((INTERNAL-READ-LIST (X)
               (IF (EQ X 'CLOSING-PARENTHESIS) NIL
                  (CONS X (INTERNAL-READ-LIST (STAT-READ))))))
      (INTERNAL-READ-LIST (STAT-READ))))
```

When we call function READ-ATOM, we know that we will find an atom at the head of the buffer; a call to the standard READ function will therefore read *only* this atom. All we have to do is to read it and to increment its private 'USE'-counter.

```
(DEFUN READ-ATOM ()
   (LET ((X (READ))) (COND
         ((NUMBERP X) X)
         (T (SETF (GET X 'USE)
                  (IF (GET X 'USE) (1+ (GET X 'USE)) 1)) X))))
```

We now have a function which reads any LISP expression and which counts the number of occurrences of atoms.

Here finally is the LISP function which acts as engine for our micro-interpreter:

```
(DEFUN LISP ()
   (PRINC "an expression please: ")
   (LABELS ((INTERNAL-LISP (X)
                (IF (NOT (EQ X 'CLOSING-PARENTHESIS))
                         (PRINT (EVAL X)))
                (FORMAT T "~% an expression: ")
                (INTERNAL-LISP (STAT-READ))))
            (INTERNAL-LISP (STAT-READ))))
```

Below is a little Le_LISP function which prints out statistics. This function uses Le_LISP's standard function OBLIST, which returns a list of all the atoms known to the LISP system. This list of all the atoms is called the *oblist*.[4] The function STAT therefore takes one element after the other from the oblist, looks to see if the indicator USE is in the P-list and, if it is, prints the name of the atom, followed by the number of times it has been used.

```
(DE STAT ()
    (LETN INTERNAL-STAT ((X (OBLIST)))
          (IF (NULL X) ()
              (IF (GET (CAR X) 'USE)
                        (PRINT (CAR X)(GET (CAR X) 'USE)))
              (INTERNAL-STAT (CDR X)))))
```

In Common-LISP there doesn't exist a function OBLIST (once you are more advanced in Common-LISP programming, try to define one for yourself!). So, to get the statistics in our Common-LISP version of the program, we will slightly modify the function, so as to allow the objects, which we want statistics of, to be given as arguments. So here is a little Common-LISP function which prints our statistics. This function STAT takes as arguments the different atoms we want to have statistics about. Therefore it takes one atom after the other from this list, looks to see if the indicator USE is on

[4] The oblist may be regarded as a sort of LISP micro-manual, accessible on-line.

he P-list and, if it is, prints the name of the atom, followed by the number of times t has been used. Note, that in order to ignore the occurrences of the different atoms given as arguments, we decrease the associated counters. This is necessary because we call the function STAT inside our private little LISP interpreter which counts all the atoms read: we don't want to count the occurrences inside the calls to STAT.

```
(DEFUN STAT (&REST SOME-OBJECTS) (COND
    ((NULL SOME-OBJECTS) ())
    (T (SETF (GET (CAR SOME-OBJECTS) 'USE)
                        (1- (GET (CAR SOME-OBJECTS) 'USE)))
       (FORMAT T "~%~A ~A" (CAR SOME-OBJECTS)
                           (GET (CAR SOME-OBJECTS) 'USE))
       (APPLY 'STAT (CDR SOME-OBJECTS)))))
```

Let us now test this new function. Here is the start of a possible interaction:

?(LISP)	*execute the interpreter function*
an expression:?1	*evaluate the number 1*
1	*the value of the evaluation*
an expression:?(Quote a)	*a second expression*
a	*its value*
an expression:? 'a	*a third expression*
<an error message>	*its value?*
What has happened?	

This is very complicated. Let us take a look: it is clearly due to the QUOTE character. Remember what we were saying in chapter 2: we stressed the fact that the character ' is simply a convenient abbreviation for a call to the function QUOTE, and if you submitted the following expression to the LISP machine

```
'a
```

it would be exactly the same as if you had entered

```
(QUOTE a)
```

It is, in fact, the latter form that the LISP reader will register. This conversion takes place at the read level itself. Characters which modify the behaviour of the LISP reader are called *macro-characters*. We shall consider them in more detail in the next section.

What has happened here is the following:

1. The LISP function started by calling function STAT-READ.

2. Function STAT-READ tests the characters at the head of the input buffer (here

the string " 'a") for occurrences of special characters such as ')', '(', '<SPACE>', '<TAB>' or '<RETURN>'. As none of these characters is present, STAT-READ calls function READ-ATOM.

3. Function READ-ATOM starts with a call to the standard read function, in the line:

```
(LET ((X (READ)))
```

From what we have already learned, we know that the value of variable X will be the list

```
(QUOTE a)
```

This list will be the first argument of the PUT function. But we were sure that the value of X would be an atom; here it is a list. Furthermore, the PUT function expects an argument of the atom type. What does the PUT function do? It generates an error.

Correcting this error involves inserting a special line into STAT-READ to handle the macro-character " ' "; this gives us the following version:

```
(DEFUN STAT-READ ()
   (LET ((X (PEEK-CHAR))) (COND
       ((EQUAL X #\( ) (READ-CHAR) (READ-LIST))
       ((EQUAL X #\) ) (READ-CHAR) 'CLOSING-PARENTHESIS)
       ((EQUAL X #\' ) (READ-CHAR) (LIST 'QUOTE (STAT-READ)))
       ((MEMBER X '(#\Space #\Newline #\Return #\Tab))
                (READ-CHAR) (STAT-READ))
       (T (READ-ATOM)))))
```

Here is an example of interaction with the new version of our LISP micro-interpreter:

```
?(LISP)                                          restart the machine
an expression:? 1                                a first expression to evaluate
1                                                its value
an expression:? '2                               a second expression
2                                                its value
an expression:? 'a                               another quoted expression
a                                                looks like it's working
an expression:? (CAR '(a b c))
a
an expression:? (CONS '(a b) '(c d))
((a b) c d)
an expression:? (STAT CAR CONS a b c d STAT)     let us see whether it counts
CAR 1                                            the number of occurrences of
CONS 1                                           each of the atoms correctly
a 3
```

```
b 2
c 2
d 1
STAT 1
nil
an expression:? (CAR '(a b c))                    take it a little further
a
an expression:? (CAR (CAR '((a b c))))
a
an expression:? (STAT CAR CONS d a b c STAT)      some more statistics
CAR 4
CONS 1
d 1
a 5
b 4
c 4
STAT 2
nil
...
```

Clearly, the real LISP reader is considerably more complicated (we shall build one in a later volume), but this model gives a first approximation to it. Look carefully at the role played by the various separator characters, as well as at the activity of the way the quote character works.

13.3 THE VARIOUS TYPES OF CHARACTER

13.3.1 Determination of the type of character

Clearly, the characters in the ASCII set do not all have the same effect during reading. Most are considered to be constituent characters of the names of atoms. On the other hand, parentheses, the space character <SPACE>, the tab character <TAB>, the return character <RETURN>, the semi-colon, the double-quote and the quote character all have special roles.

We shall start by summarizing what we already know about the role of these different characters:

<SPACE> <TAB>,<RETURN>	These characters are separators. These characters cannot normally be included in the names of atoms; they are used to separate sequences of atoms.

`;`

The semi-colon is the start of comment character. The <RETURN> character is an *end of comment* character. If you want to put comments in your program, they must start with a semi-colon. They end with a <RETURN>. The standard LISP reader ignores all comments.

`"`

The double quote character acts as delimiter for strings of characters. Within a character string, no checking takes place on the role played by characters. On the other hand, as in language C, there is a convention in LISP allowing unusual characters to be included in strings. For this it is sufficient to precede the special character with a '\' (backslash). Naturally, the only special character in a string is the double quote (") and the backslash itself. So, for example, to include a double quote in a string, we have to write

```
"ab\"ba"
```

to get the string ab"ba.

`(,)`

Clearly, the opening and closing parenthesis characters mark the beginning and end of lists. These two characters are also separators and cannot therefore normally be used within atom names.

`'`

The quote character tells the LISP reader that it should generate a call to function QUOTE with the element immediately following the character as argument. Characters modifying the behaviour of the LISP reader are called *macro-characters*.

`\`

The backslash character is used to quote special characters, so that these characters temporarily lose their meaning.

`#`

We have already seen the use of the sharp sign in combination with the backslash (\) to name character objects, and its use in combination with the quote-character (') to quote functions. The sharp sign is a very special character which exists in combination with other characters, which we will see later.

`|`

The vertical bar character is, like the double quotes, always used in pairs to delimit names which may contain special characters. The difference between the vertical bar and the double quotes is that the former delimits strings denoting an atom or, as we say in LISP, interned symbols. The latter, the double quotes, are always delimiting strings.

As an example for the use of the vertical bar look at the following example:

```
(SETQ \(\a\n\ \a\t\o\,m\'\s\ \v\a\l\u\e\) 3)
```

This creates an atom called "(an atom's value)" and gives it the value 3. In order to create this very special symbol we had to quote each character, the normal characters were quoted in order to prevent their transformation in upper case letters in Common-LISP, and the special characters (,) and ' were quoted in order to prevent the LISP reader from interpreting these characters. Surely, this is not easy to read. We can do exactly the same by writing:

```
(setq |(an atom's value)| 3)
```

where the two delimiting vertical bars tell the LISP reader not to interpret the special characters and not to transform lower case letters in upper case letters. Definitely, this is – for us – much easier to read.

In Le_LISP, if at any moment during your introduction to LISP, you can no longer remember the role played by a character, you can call function TYPECH which gives you the TYPE of the CHaracter. Here is the definition of the function:

(TYPECH *e*) → *type*

argument *e* may be a character string (of length one character) or a single character atom. The *type* is the current type of that character.

(TYPECH *e type*) → *type*

the *type* argument must be a character type. The *type* returned is the new type of character *e*.

In Le_LISP the type of a character is defined as follows:

type	*character*
CNULL	ignored characters
CBCOM	start of comment
CECOM	end of comment
CQUOTE	QUOTE character in Le_LISP
CLPAR	opening parenthesis
CRPAR	closing parenthesis
CDOT	dot character
CSEP	separator

CMACRO	macro-character
CSTRING	string delimiter
CPNAME	normal character
CSYMB	special symbol delimiter
CPKGC	package delimiter
CSPLICE	splice-macro
CMSYMB	single character symbol

With the knowledge you already have and this new function, you can write some funny things. For example, if you do not like parentheses, after the instructions

```
(TYPECH "{" (TYPECH "("))
(TYPECH "}" (TYPECH ")"))
```

you will be able to write either

```
(CAR '(DO RE MI))
```

or

```
{CAR '{DO RE MI}}
```

In Common-LISP, the typing of characters is lightly different: Common-LISP distinguishes between *illegal, whitespace, constituent, single escape, multiple escape* and *macro*-characters. Macro-characters are further divided into the types *terminating* and *non-terminating*.

Here is a list of the standard character syntax types:

whitespace	return, tab, space, page, newline, linefeed
single escape	\
multiple escape	|
terminating macro	", ', `, (,), ;
non-terminating macro	#
constituent	all other characters

Constituent means that the corresponding character may be used in the name of a symbol without any problem. All other characters must be *escaped* if one wants to use them as part of the print-name of a symbol. One can *escape* a *single* character with the backslash; one can *escape multiple* characters with the vertical bar. *Whitespace* characters are separators, they end the name of a symbol, otherwise they don't have any meaning. *Terminating macro*-characters also end the name of a symbol, but they are interpreted by LISP as having an additional meaning. *Non-terminating macro*-characters may be part of the name of a symbol; one doesn't need to escape them.

In Common-LISP, there is no function, like the Le_LISP function TYPECH, which gives the type of a character. The best is to look up a manual. But for changing the syntax of a character, there is a function *almost* like TYPECH: this is the function SET-SYNTAX-FROM-CHAR. Here is its definition:

(SET-SYNTAX-FROM-CHAR *to-char from-char*)

This function takes two characters as argument: *to-char* and *from-char*. Its effect is to give *to-char* the same syntactic definition as has the character *from-char*.

For example, evaluation of

(SET-SYNTAX-FROM-CHAR #\^ #\')

gives the character ^ the same syntax *and* the same functionality as has the quote-character. After this we can as well write

(QUOTE A)

as

'A

as

^A

all three will have the same effect.

Normally, this works very well with user-defined special (macro) characters, be careful when handling standard macro-characters: the effect may be different on different versions of Common-LISP. Especially if the functionality of the closing parenthesis is too much hidden in the implementation of the LISP reader, there is no easy way – in Common-LISP – to transfer its functionality to another character. Here again, the best is to look up the Common-LISP manual; we won't talk about it here: it is too complicated for a first introduction.

13.3.2 Macro-characters

Macro-characters modify the behaviour of the LISP reader to generate function calls. So the quote-character, the only macro-character that you know so far, translates

'SOMETHING

into

(QUOTE SOMETHING)

The LISP user can define personal macro-characters using a special functions to define new macro-characters. In Le_LISP this function is called DMC, in Common-LISP the corresponding function is SET-MACRO-CHARACTER. Since there is some difference in the way these two LISPs handle the definition of macro-characters, we will first show how it is done in Le_LISP (or similar dialects of LISP) and then how it is done in Common-LISP (or MACLISP).

13.3.2.1 Definition of Macro-Characters in Le_LISP

The function DMC, Le_LISP's function for defining new macro-characters, can be defined syntactically as follows:

```
(DMC c () function-body)
```

The first argument, *c*, must be a one character symbol, preferably written surrounded by a pair of vertical bars. When the LISP-reader meets this character, it will evaluate the *function-body*. Finally, the result of the evaluation will replace the macro-character in the input stream.

For example, if the quote-character had not been defined already, we could define it as follows:

```
(DMC |'| () (LIST (QUOTE QUOTE)(READ)))
```

This definition can be read as a command to the LISP reader saying that each time it meets a character ' in the input stream it must replace that character by a list whose first element is the atom QUOTE and the second element is the LISP expression immediately following the character.

Let us consider another example: we shall construct macro-characters allowing us to write

[*argument*$_1$ *argument*$_2$ . . . *argument*$_n$]

instead of

(LIST *argument*$_1$ *argument*$_2$ . . . *argument*$_n$)

To do so, we have to define two macro-characters: the macro-character "[" which must construct the call to function LIST with as many arguments as there are LISP expressions up to the following "]" character. The latter character must also be a macro-character, if only to act as a separator. Here are the two definitions:

```
(DMC |[| ()
        (LETN SELF ((X 'LIST))
        (AND (NEQUAL X "]") (CONS X (SELF (READ))))))

(DMC |]| () "]")
```

The function NEQUAL is obviously an abbreviation of (NULL (EQUAL))[5] giving us the following relation:

[5] If this function does not exist in your version of LISP, you can easily define it using:

```
(DE NEQUAL (X Y) (NULL (EQUAL X Y)))
```

(NEQUAL x y) ≡ (NULL (EQUAL x y))

As you can see, the definitions of macro-characters may be of any complexity. Their role is always to simplify the writing of programs.

Here are a few examples of the use of these macro-characters:

```
[1 2 3)                          → ( 1 2 3)
['a 'b (CAR ' (1 2 3 ) 'c]       → (a b 1 c)
'[1 2[3 4] 4]                       (LIST 1 2 (LIST 3 4) 4)
(CONS 1[2 3 4])                     (1 2 3 4)
```

You can now use the macro-character quote (') to enter lists and the macro-characters '[' and '[' to generate calls to LIST.

13.3.2.2 Definition of Macro-Characters in Common-LISP

In this paragraph, let us see how to define macro-characters in Common-LISP. The Common-LISP function to define macro-characters is called SET-MACRO-CHARACTER. It is syntactically defined as:

(SET-MACRO-CHARACTER *char function*))

Evaluation of this function causes *char*, a character, to be a macro-character that, when seen by READ causes *function* to be called. Clearly, SET-MACRO-CHARACTER is a function associating a function to a character and changing the type of the character, like the function DMC of Le_LISP. The difference with DMC is 1) the argument is a character object, and 2) the second argument is a function.

Suppose again that the macro-character ' is not defined. We could define this macro-character by attaching the following function to the quote-character ':

```
(DEFINE MY-QUOTE (STREAM CHAR)
        (LIST (QUOTE QUOTE) (READ STREAM T NIL T)))
```

Let us examine this function in some detail. First we note that this is a function with two arguments, STREAM and CHAR. When this function is called, that is when the LISP reader encounters the quote-character, the first argument receives a stream. We will see later in this chapter how to handle streams. At this point, just note that all reading is done through a stream, which may be connected to a terminal or a file (or even to other LISP objects). By default, the stream on which one reads is the terminal. Normally one doesn't have to specify it. It is the function READ which will automatically transmit its stream to the functions associated to macro-characters.

The second argument transmitted to functions associated to macro-characters, is the macro-character itself. In this example, we don't use this argument at all.

The function MY-QUOTE tells that, when it is called with a stream and a character, it should return a list composed of two elements: the symbol QUOTE and the LISP expression following this symbol in the input stream. Note that the function READ here has four arguments. This changes what we had said previously about it: it is a function with up to four optional arguments. If no argument is given, READ reads one expression of the current input stream. The first argument may indicate a stream. Here, since anyway we have the actual input stream available, we give this first argument. The second and third argument tells READ what to do when an *end-of-file* is encountered. For the moment, just note that we give these two arguments always in this form in a call to READ inside a function associated to a macro-character. The fourth argument has to be T if we call read from a function associated to a macro-character. This argument is necessary (for obscure implementation reasons) whenever there is a *recursive* call to READ. Recursive here means that we call READ from within READ. Naturally, since we want to evaluate calls to this function during the reading of LISP expressions.

All that is missing still, is to associate this function MY-QUOTE to the quote-character ' and to do this, we use SET-MACRO-CHARACTER as follows:

```
(SET-MACRO-CHARACTER #\' #'MY-QUOTE)
```

A call to SET-MACRO-CHARACTER always returns T.

Let us rewrite in Common-LISP the macro-characters [and], permitting us to use

[*argument*$_1$ *argument*$_2$ . . . *argument*$_n$]

as a shorthand for

(LIST *argument*$_1$ *argument*$_2$ . . . *argument*$_n$)

To do so we will, as previously, write a function for the macro-character [which must construct the call to LIST and read all expressions until the next occurrence of].

Here is a Common-LISP definition for this macro-character:

```
(DEFUN OPENING-[ (STREAM CHAR)
        (LABELS ((MY-LOOP (X)
                   (AND (NOT (EQUAL X '|]|))
                        (CONS X (MY-LOOP
                                  (READ STREAM T NIL T))))))
            (MY-LOOP 'LIST)))

(SET-MACRO-CHARACTER #\[ #'OPENING-[)
```

Note again that we gave two arguments to this function, one for the current stream, one for the macro-character. Note also the recursive call to READ. Your calls to READ within a macro-character should invariably have this form.

There is still one problem: such as we have defined the macro-character [, it will work for examples like:

```
[1 2 3 ]
```

or

```
['a 'b 'c ]
```

but not for calls like:

```
[1 2 3]
```

or

```
['a 'b 'c]
```

In the two second examples, the closing bracket] would not be recognized as a special character, but it would just be part of the atom 3] or C]. Until now, it is just a character like every other normal character: it is a *constituent* character. In order to make it a separator, we have to transform it into a macro-character. Here is its definition:

```
(DEFUN CLOSING-] (STREAM CHAR) '|]|)

(SET-MACRO-CHARACTER #\] #'CLOSING-])
```

All it does is to return a symbol which may be recognized by the macro-character [as a corresponding closing bracket.

At the end of the previous paragraphs are some examples of the use of these macro-characters.

13.3.2.3 A More Elaborate Example

Let us now build a final set of macro-characters and define the macro-character *back-quote* (`). This very useful macro-character is provided in most LISP systems. It is in a sense a combination of the characters ', [and]. It can be used when we need a long quoted list to contain an element that has to be evaluated; in other words, it is used in a list where most elements are quoted.

Let us take an example. Assume that variable X is bound to a list (D E) and that we want to construct a list with the following elements in the following order: A, B, C, the value of X, and the element F. The only way we can currently do that is to write

```
(LIST 'A 'B 'C X 'F)
```

or

```
['A 'B 'C X 'F]
```

If we now want to use these elements to construct the list:

```
(A B C D E F)
```

we have to write:

```
(CONS 'A (CONS 'B (CONS 'C (APPEND X (LIST 'F)))))
```

Finally, if we want to use the same elements to construct the list:

```
(A B C E F)
```

we have to write:

```
(CONS 'A (CONS 'B (CONS 'C (CONS (CADR X)(LIST 'F)))))
```

These are cumbersome expressions, given that only one element in the list has to be evaluated and all the others are given as they stand.

The macro-character *backquote* gives us a typographic device making it possible to write these forms much more easily. Here is how to construct the three lists given as examples, using this new macro-character:

```
`(A B C ,X F)            →  (A B C (D E) F)
`(A B C ,@X F)           →  (A B C D E F)
`(A B C ,(CADR X) F)     →  (A B C E F)
```

Clearly, the macro-character *back-quote* uses two further macro-characters, the comma (,) and the atsign (@). Within a *backquoted* list an expression preceded by a comma gives the value that is then inserted as an *element* in the final result. An expression preceded by the characters , and @ gives a value later inserted as a *segment* into the final result. All other expressions are taken as they stand. The results of a call to function QUOTE and a call to function BACKQUOTE are therefore identical if there are no commas or atsigns within the backquoted expression.

Now all we have to do is to define this macro-character, for which we have to use, in Le_LISP, the following very complicated form:

```
(DMC|`|()
  (LET ((COMMA (TYPCH ",")))
       (TYPECH "," (TYPCH " "))
       (LET ((QUASI ['EVAL ['BACKQUOTE ['QUOTE (READ)]]]))
            (TYPECH "," COMMA)
            QUASI)))
```

What we have done here is two things. We started by limiting the range of the macro-character"," (comma). We saved its character type on entry to the definition, and then set its type to the value of the macro-character quote, to ensure that within the range of the macro-character ' (backquote) the macro-character , (comma) is defined as we want it to be. On exiting from backquote we restore its original type, using the instruction

```
(TYPECH "," COMMA)
```

We cheated slightly in our temporary definition of the macro-character, , since, in Le_LISP, TYPECH does not really change the definition associated to a macro-character, but just its type of character. Mainly, what we did is to keep the old character type of , in the variable COMMA, then forcing the type of macro-character on , , and finally restoring the old type of this character. In all this we did not touch the function which may be associated to this character: we just assumed that if the character comma isn't a macro-character outside the backquote, it remains outside the backquote as a standard character.

In the Common-LISP definition of the same macro-character, we change the function definition associated to the character , inside the backquote, and we restore its old function definition before exiting the backquote macro-character. This we do in saving the function definition associated to , in a variable, and resetting the character to this saved function in a call to SET-MACRO-CHARACTER. The function GET-MACRO-CHARACTER takes as argument a character and returns the associated function definition, if one exists, or NIL if there doesn't exist a function definition.

So, finally, here is the Common-LISP definition of the macro-character backquote:

```
(DEFUN |BQ-`| (STREAM CHAR)
   (LET ((COMMA  (GET-MACRO-CHARACTER  #\,)))
        (SET-MACRO-CHARACTER  #\,  #'INTERNAL-COMMA)
       LET
        ((QUASI
            ['EVAL
                 ['BACKQUOTE ['QUOTE (READ STREAM T NIL T)]]]))
        (SET-MACRO-CHARACTER  #\,  COMMA)
        QUASI)))

(SET-MACRO-CHARACTER  #\`  #'IBQ-`|)
```

Next, we called a function called BACKQUOTE to carry out the activity we expect from the backquote. This function will have to build up calls to functions LIST and CONS as necessary to construct the correct result list. Clearly, in order actually to build that list, we have to evaluate the calls generated, which we do with an explicit call to function EVAL.

Let us now look at the BACKQUOTE function:

```
(DEFUN BACKQUOTE (EXPR) (COND
   ((NULL EXPR) NIL)
   ((ATOM EXPR) ['QUOTE EXPR])
   ((EQ (CAR EXPR) '*UNQUOTE*) (CADR EXPR))
   ((AND (CONSP (CAR EXPR)) (EQ (CAAR EXPR) '*SPLICE-UNQUOTE*))
       ['APPEND (CADAR EXPR) (BACKQUOTE (CDR EXPR))])
   ((COMBINE-EXPRS (BACKQUOTE (CAR EXPR))
               (BACKQUOTE (CDR EXPR)) EXPR))))
```

BACKQUOTE receives as argument the expression which immediately follows the character ` (backquote). This expression is associated with the variable EXPR in order to deal with the five cases that may occur:

1. Expression EXPR is empty: do nothing.

2. The expression EXPR is an atom: BACKQUOTE generates a call to function QUOTE;
   e.g.: if the value of EXPR is *atom*, BACKQUOTE generates: (QUOTE *atom*)

3. The expression starts with the special atom *UNQUOTE*: this atom was generated by the character , (comma); it is therefore the value of the argument of , that has to be returned. This is done by replacing the expression (*UNQUOTE* *arg*) in the expression read, by arg in the expression generated.
   e.g.: if the value of EXPR is (*UNQUOTE* *argument*), BACKQUOTE generates:

   argument

4. The first element in EXPR is a list beginning with the special atom *SPLICE-UNQUOTE*: this atom has been generated by the characters, ,@; this means we have to concatenate the value of the argument of *SPLICE-UNQUOTE* to the end of the list. This is done by generating a call to function APPEND whose first argument is the argument of *SPLICE-UNQUOTE* and whose second argument is the result of the recursive call to BACKQUOTE with the CDR of the expression EXPR.
   e.g.: if the value of EXPR is ((*SPLICE-UNQUOTE* *argument*) *rest-of- arguments*), BACKQUOTE generates:

   (APPEND *argument* (BACKQUOTE *rest-of-arguments*))

5. In all other cases, the results of the analysis of the CAR and the CDR of EXPR must be combined. Four cases have to be distinguished:

a. The CAR and the CDR are constants: in this case we only generate a call to function QUOTE with the whole expression.
b. The CDR is NIL, in which case we have to generate a call to function LIST with the CAR as argument.
c. The CDR is a list which begins with a call to LIST: we know then that this call has been generated by BACKQUOTE and we therefore insert the CAR as a new first argument of the call.
d. In all other cases, we generate the call to function CONS with the CAR as the first argument and the CDR as the second argument, which has the effect of restoring the original list.

Here is the function COMBINE-EXPRS:

```
(DEFUN COMBINE-EXPRS (LFT RGT EXPR) (COND
        ((AND (ISCONST LFT)(ISCONST RGT)) ['QUOTE EXPR])
        ((NULL RGT)['LIST LFT])
        ((AND (CONSP RGT)(EQ(CAR RGT) 'LIST))
            (CONS 'LIST (CONS LFT (CDR RGT))))
        (['CONS LFT RGT])))
```

which uses the auxiliary function ISCONST:

```
(DEFUN ISCONST (X)
   (OR (NULL X)(EQ X T)(NUMBERP X)(CHARACTERP X)
       (AND (CONSP X)(EQ (CAR X) 'QUOTE))))
```

Finally, all we need do is to define the macro-character, (comma). Since we want this macro-character to apply only within the BACKQUOTE, we have to take care that the definition of the macro-character does not *definitively* modify its type. This is why, in Le-LISP, we put the DMC within a LET which simply saves and restores the type of character, before and after its definition as a macro-character. Here then is the program sequence:

```
(LET((COMMA (TYPECH ",")))
    (DMC ","()
         (LET ((X (PEEKCH)))
                      (IF (NULL (EQUAL X "@"))['*UNQUOTE*(READ)]
                         (READCH)
                         ['*SPLICE-UNQUOTE* (READ)])))
       (TYPECH "," COMMA))
```

Note in particular the use made of function PEEKCH.

And here the corresponding version of the function in Common-LISP. Remember that we just define the associated function, the association between this function and the character , will only be done inside the backquote.

```
(DEFUN INTERNAL-COMMA (STREAM CHAR)
   (LET ((X (PEEK-CHAR STREAM T NIL T))) (COND
        (NULL (EQUAL X #\@))['*UNQUOTE* (READ STREAM T NIL T)])
        (T (READ-CHAR STREAM T NIL T)
           ['*SPLICE-UNQUOTE* (READ STREAM T NIL T)]))))
```

Let us now examine a few examples of this new macro-character. By way of a first example let us assume to give LISP the expression:

```
`(A B C)
```

since all the elements in this list are constants, the function BACKQUOTE generates the call:

```
(QUOTE (A B C))
```

and its evaluation therefore gives the list:

```
(A B C)
```

Returning to the examples above and assuming that variable X is linked with list (D E) and variable Y with the atom F, if we submit the following expression to LISP:

```
`(A B C ,X Y)
```

the LISP reader will translate it into:

```
(EVAL (BACKQUOTE (QUOTE (A B C (*UNQUOTE* X) Y))))
```

which, after evaluation of the call to BACKQUOTE, will give us:

```
(EVAL '(CONS (QUOTE A)
     (CONS (QUOTE B)
          (CONS (QUOTE C)
               (CONS X (QUOTE (Y)))))))
```

whose evaluation will finally give us:

```
(A B C (D E) Y)
```

For our second example, take the expression:

```
`(A B C ,@X ,Y)
```

which will be translated by the LISP reader into:

```
(EVAL (BACKQUOTE
    (QUOTE (A B C
        (*SPLICE-UNQUOTE* X)
        (*UNQUOTE* Y)))))
```

After evaluation of the argument of EVAL, this will give:

```
(EVAL '(CONS (QUOTE A)
    (CONS (QUOTE B)
        (CONS (QUOTE C)
            (APEND X (LIST Y))))))
```

which returns the list:

```
(A B C D E F)
```

Let us stop here in our examination of input macro-characters (we shall return to the subject) and briefly look at how to modify an input-output file.

13.4 HOW TO REDIRECT INPUT/OUTPUT

So far, we have only been able to enter LISP functions through the terminal keyboard, and display the results of computations on the terminal screen. But reading and printing can involve files and devices, not just the terminal. For example, we often need to prepare programs under an editor and to load them into LISP afterwards, or to save results in files to be able to use them later. This section will show you how to carry out this kind of operation.

LISP provides a function to read in the content of files. This function is called LOAD. It is very useful for *loading* a program which is in a file. Here is the definition of LOAD:

(LOAD *filename*) → T
and reads the file called *filename* from beginning to end, evaluating each of the expressions contained therein.

Consequently, if you are working, for example, under a UNIX system, and if the program of the preceding section (the definition of *backquote*) is in the file backquote. lisp in a subdirectory my-lisp, the command

```
(LOAD "my-LISP/backquote.LISP")
```

will read the file, definition by definition.

The argument *filename* is a string of characters corresponding to the way in which the file might be specified in your operating system. For example, if, you want to load the UNIX file /usr/hw/lisp/example.lisp, it can be specified by using the string "/usr/hw/lisp/example.lisp". In MULTICS, the comparable file specification would be ">udd>hw>lisp>example.lisp", or, under some strange UNIX like system it might be "\hw\lisp\example.lisp".

This function is probably the most used input/output function handling files: all the programs you have saved in files have to be loaded before being executed.

There are still other uses for redirecting input/outputs: you may want to print results or partial results into files or you may want to read data from a file. Common-LISP handles all these possibilities in giving to the input/output functions an additional argument. This argument should be a *stream*. A stream is a special data type used for input/output processing, and serves as a source (for input operations) or sink (for output operations). Normally streams are connected to files, terminals or other devices. They are the interface between LISP and the operating system. So, in order to do input/output operations to a file, one has first to set up a stream that refers to that file. This is done with the LISP function OPEN. Like LOAD, the function OPEN expects a *filename* as argument. In addition one has to specify if one wants to read from the file, to write from the file, or if one just wants to know if the file exists. The result of a call to OPEN will be the setting up of a stream interfacing LISP with this file. It is this stream which is returned.

Let's take an example: to open the file test.lisp for output operations, we have to call OPEN as follows:

```
(OPEN "test.lisp" :DIRECTION :OUTPUT)
```

This will return something like:

```
#<output stream "test.lisp">
```

indicating that a stream to the specified file has been opened. Actually, what the machine prints out exactly, may vary from one implementation of LISP to the other. But, in all cases, the value returned is a stream.

To use this stream in actual output operations, we should keep it somewhere. So it is better to write something like:

```
(LET ((MY-OUTPUT-STREAM (OPEN "test.lisp" :DIRECTION :OUTPUT))) . . .
```

Then, later on, we can use the variable MY-OUTPUT-STREAM to refer to the stream just opened.

Once we have finished printing to this file, we should *close* the stream connecting to it. This we do with the function CLOSE, which expects a stream as an argument. Its effect is to close the existing connection between LISP and the file. Here, for example, once we've finished writing to the file test.lisp, we close the connection with:

```
(CLOSE MY-OUTPUT-STREAM)
```

Naturally, you can also create a stream for input. So if you intend to read data from the file data.lisp in your current directory, you have to give the command:

```
(OPEN "data.lisp" :DIRECTION :INPUT)
```

which will return a stream, such as:

```
#<input stream "data.lisp">
```

usable during future input operations. One can also open a file for reading and writing, that is: one can create bidirectional streams through the direction :IO. So the call

```
(OPEN "my-file" :DIRECTION :IO)
```

returns a stream which may be used in input and output operations. There exists one more possible direction while opening a stream: this is :PROBE. If one gives this direction as argument to a call to OPEN, Common-LISP does not really create a stream, but returns just NIL if the file given as argument doesn't exist, and something different from NIL if the file exists. So, the call:

```
(OPEN "my-test-file" :DIRECTION :PROBE)
```

returns #<probe stream "my-test-file"> if the file my-test-file exists and NIL otherwise.

Note that OPENing a non existent file for :OUTPUT or :IO creates a file with the name given as an argument to OPEN. If one tries to OPEN a non existent file for :INPUT, LISP gives an error message.

OK, so far we know how to create streams for reading, writing or both. To write something in an output stream we use the function PRINT with a stream as second argument. Here is an example:

```
? (DEFUN ALL-CDR-REVERSE (L STREAM)
      (AND L (PRINT (REVERSE L) STREAM)
             (ALL-CDR-REVERSE (CDR L) STREAM)))


? ((LET ((X (OPEN "my-file" :DIRECTION :OUTPUT)))
       (ALL-CDR-REVERSE '(A B C) X)
       (CLOSE X))
  = T
```

In this example, we first defined a function which prints the reverse of the successive cdr's of a list. Note that the PRINT function call has two arguments: a list to print and a stream to print to.

The LET does three things:

- First it opens an output stream to the file my-file. This stream is kept in the variable X.
- Second, it calls ALL-CDR-REVERSE with the list (A B C) and the stream X.
- Third, it closes the stream X, which breaks the connection to the file my-file.

The last line in this example is the value returned by the call to LET, the atom T, which in turn is the value returned by CLOSE.

As you can see, nothing of all the calls to PRINT was printed on the terminal, because all the output went to the file my-file. Let us read back the content of this file to see what is in it:

```
? (LET ((X (OPEN "my-file" :DIRECTION :INPUT)))
        (PRINT (READ X))
        (PRINT (READ X))
        (PRINT (READ X))
        (CLOSE X))
(C B A)
(C B)
(C)
= T
```

What we did here is: we first opened the file for input creating a stream which we kept in the variable X. Then we printed what we read on this stream. Note that the call to PRINT here has just one argument. If this is the case, or if the second argument, the stream, is NIL, LISP redirects the output to the terminal. READ has one argument, the stream we want to read from.

All input/output functions conform to this general scheme: they all accept a stream as an additional argument. If this argument is missing, or if it equals NIL the input or output is directed towards the terminal (or, more precisely: towards the stream indicated in the global system variables *STANDARD-INPUT* and *STANDARD-OUTPUT*), otherwise, the input or output is directed towards the stream given as an argument.

This rule explains also the first argument T we gave in all examples of the formatting function FORMAT. The T as first argument to these functions indicates that we want to print to the standard output. If we want to print to another stream, we just have to give this other stream as an argument in place of T.

One last remark about the function READ. In the previous section we saw the calls to read with four arguments, such as:

```
(READ STREAM T NIL T)
```

Clearly, the first argument is the stream from which we want to read. We also know that the last argument indicates that this call to READ takes place during another READ, it's the *recursive-p* argument for calls to READ inside macro-characters. There remains still the second and third argument. These last two tell us what to do when READ meets the end of a file. If the second argument is T, READ will give an error message when an attempt is made to read further than the end of a file. If this argument is NIL, then the third argument indicates the value of READ if an attempt is made to read further than an end of file. For example, the following call:

```
(READ MY-INPUT-STREAM NIL '|The End|)
```

would normally read an expression if there is still one available, and if there is nothing more to read, this call would return the atom |The End|. This is very useful if one wants to read all expressions of a file and one doesn't know a priori their number.

Let us take an example, and assume that we have a file called values, containing the following list of atoms:

```
1
2
3
4
5
6
7
8
9
10
```

and let us build a function which computes the sum of all the numbers in a file. Here it is:

```
(DEFUN SUM-OF-NUMBERS (FILE)
    (LABELS ((AUX-SUM (NUMBER STREAM) (COND
                        ((EQUAL NUMBER '|The End|) 0)
                        ((NUMBERP NUMBER)
                            (+ NUMBER (AUX-SUM
                            (READ STREAM NIL '|The End|)STREAM)))
                        (T (AUX-SUM
                          (READ STREAM NIL '|The End|)STREAM)))))
          (LET ((MY-STREAM (OPEN FILE :DIRECTION :INPUT)))
            (LET ((RESULT (AUX-SUM 0 MY-STREAM)))
              (CLOSE MY-STREAM)
              RESULT))))
```

Note that, since we don't know the number of objects in the file, we use the value returned by READ, when reaching the end of the file, as recursion stop test.

To calculate the sum of the numbers contained in the file values, we simply have to call this function as follows:

```
(SUM-OF-NUMBERS "values")
```

Notes on Le_LISP input/output

To close this chapter, we will have a short look at input/output functions of Le_LISP.

The Common-LISP function LOAD is called LOADFILE in Le_LISP. It behaves exactly the same.

In Le_LISP, instead of having one function with a lot of specifiers, like the Common-LISP OPEN, we have a set of different functions for the different roles of OPEN. To open a file for input, Le_LISP offers the function OPENI. It takes one argument, the name of the file one wants to read from, and returns a stream connecting to this file. To open a file for output, Le_LISP uses the function OPENO with the file name as argument. This function too returns a stream. Finally, to test the existence of a file, the Le_LISP programmer uses the function PROBEFILE, which takes a file name as argument and returns T if the named file exists, NIL otherwise. Like in Common-LISP, a stream is closed with the function CLOSE.

More profound differences exist when one wants to redirect the input or the output. Le_LISP may, like Common-LISP, have several open streams, but all output operations pass through one channel, and it is this channel which is connected to the stream. So before calling the PRINT function, one has to associate this channel to the stream with the help of the function OUTCHAN. For example, the following five instructions will

- open the file test.11
- associate the output channel to the stream connected to the file test.11
- write the message *Here I am* into the file test.11:
- re-associate the output channel with the terminal
- close the stream connected to test.11.

```
(LET ((MY-STREAM (OPENO "test.11")))
     (OUTCHAN MY-STREAM)
     (PRINT "Here I am")
     (OUTCHAN NIL)
     (CLOSE MY-STREAM))
```

The corresponding function for associating a stream with the input channel is INCHAN.

This way of handling the redirection of input/output has some advantages when one has to write a large amount of data in one and the same file: PRINT is faster since it is not responsible for testing *where* the output directs to, the Common-LISP scheme has advantages when one has often to change the directions of input and output.

Finally, in graphic form, here are the various possibilities in LISP for exchanges with the outside world:

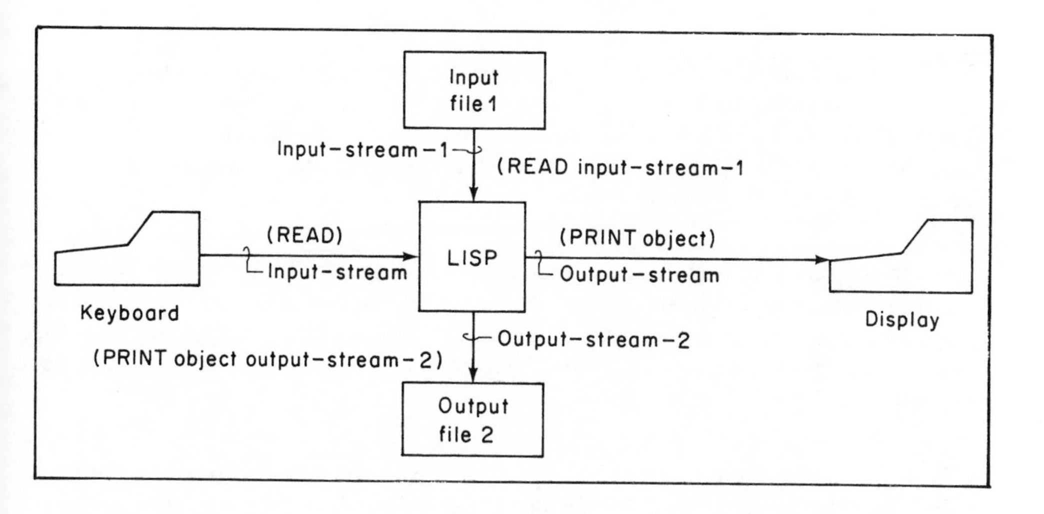

Note that in these graphical representations the arrows represent the streams which connect LISP to files or devices.

13.5 EXERCISES

1. To avoid having to write (LOAD *filename*) each time you want to load a file, write a macro-character ^ allowing you to load files using the command

   ^*filename*

2. Define the macro-characters >= and <= for GE and LE respectively.

3. a. Construct a program which would write even numbers from one to *n* and the atom END to a file called *even.no*.

   b. Construct another program which would write the odd numbers from 0 to *n* and the atom END to a file called *odd.no*.

   c. Construct a third program which writes the sums of the pairs of even and odd numbers from files *even.no* and *odd.no* to a file call *sum.no*.

14 PATTERN MATCHING (PART ONE)

When you are programming in LISP, you often have to find elements from inside a list or to compare two lists. The functions that we know so far which will help us handle this type of problem are EQUAL, which compares two lists element by element, and the function MEMQ and MEMBER which look for the occurrences of an atom or of any element respectively within a given list.

Though they are very useful, these functions are very restricted. Above all, they cannot be used to make approximate comparisons between two lists: for example, to discover whether a list includes a particular sequence of elements in a given order, or to find out whether a list contains only three elements with a particular atom as the final element.

Clearly, we could always write a specific program to solve each comparison problem independently. For example, to find out whether a list L contains the three elements DO, RE and MI in that order, we could define function DOREMI as follows:

```
(DEFUN DOREMI (L)
        (LET ((X (MEMQ 'DO L))) (COND
             ((NULL L) ())
             ((AND (EQ (CADR X) 'RE) (EQ (CADDR X) 'MI)) T))))
```

or to discover whether a list has only three elements with the final element being the atom MI, we could construct function QUICK as follows:

```
(DEFUN QUICK (L)
        (IF (AND (= (LENGTH L) 3) (EQ (CAR (REVERSE L)) 'MI)) T NIL))
```

but this type of function is likely to multiply very rapidly, with a specific function for each particular case we meet.

Pattern matching is a technique which makes it possible to compare a specific list with a list '*pattern*', where the *pattern* is a way of describing the *structure* of a list. To return to the two examples above, the corresponding patterns would be

```
(& DO RE MI &)
```

and

($ $ MI)

In these patterns, the sign & means any sequence of elements while the $ means any one element.

So patterns are visually comprehensible descriptions of list structures. The process of using a pattern to search a list is called *pattern matching*. In this chapter, we shall see how to construct a program to undertake *pattern matching* of this kind.

14.1 FIRST VERSION OF A PATTERN MATCHING FUNCTION

We shall start by considering patterns made up of *constants* alone, i.e. of random atoms and $ signs. Here are a few examples of what our comparison function MATCH might do:

```
(MATCH '(DO RE MI) '(DO RE MI))            → T
(MATCH '(DO RE MI) '(DO RE DO))            → NIL
(MATCH '(DO $ MI) '(DO RE MI))             → T
(MATCH '(DO $ MI) '(DO DO MI))             → T
(MATCH '(DO $ MI) '(DO (DO RE MI)MI))      → T
(MATCH '(DO $ MI) '(DO RE))                → NIL
(MATCH '(DO $ MI) '(DO FA RE))             → NIL
(MATCH '($ $ $) '(1 2 3))                  → T
```

The first pattern, (DO RE MI), simply compares two identical lists as neither of the special signs $ or & is used.

The second pattern, (DO $ MI), will match any three-element list starting with atom DO and ending with atom MI. The second element may be any atom or any list. The $ sign simply indicates that there must be *one* element at that point.

The third pattern, ($ $ $), will match any list containing exactly three elements.

Pattern matching, with $ as the only special sign, therefore behaves almost like the function EQUAL: if no occurrence of $ appears in the pattern, pattern matching *is identical* to the test for equality. If there is an occurrence of the $ sign the comparison can simply skip the corresponding element in the list given.

Before we write our first version of function MATCH, let us take another look at function EQUAL (cf section 6.3):

```
(DEFUN EQUAL (ARG1 ARG2) (COND
    ((ATOM ARG1) (EQ ARG1 ARG2))
    ((ATOM ARG2) NIL)
    ((EQUAL (CAR ARG1)(CAR ARG2))(EQUAL (CDR ARG1)(CDR ARG2)))))
```

Writing MATCH poses no problems:

```
(DEFUN MATCH (PATTERN DATA) (COND
   ((ATOM PATTERN) (EQ PATTERN DATA))
   ((ATOM DATA) NIL)
   ((EQ (CAR PATTERN) '$)
        (MATCH (CDR PATTERN)(CDR DATA)))
   ((MATCH (CAR PATTERN) (CAR DATA))
        (MATCH (CDR PATTERN) (CDR DATA)))))
```

This function allows pattern matching with the $ sign, and therefore the extraction of data with the same number of elements as the pattern.

If we also want to allow the use of the sign & to stand for *sequence* of elements, we would have to introduce an additional clause:

```
((EQ (CAR PATTERN) '&) . . .)
```

To find out what we have to substitute for the ". . .", start by looking at an example. The pattern:

```
(& DO &)
```

will match any list containing at least one occurrence of atom DO. Consequently, the pattern will work for the list (DO), as well as for any list where this atom is preceded by any number of elements, as, for example, in the lists

```
(1 DO)
(1 2 DO)
(1 1 2 DO)
(A B C D E F G DO)
etc
```

as well as lists where this atom is followed by any number of elements, of any kind, as in the following examples:

```
(DO 1)
(DO 1 2 3)
. . .
(1 DO 1)
(1 2 DO 1 2 3)
etc
```

The ampersand sign may therefore correspond to nothing (as is the case if the list submitted as data is (DO)), or to any number of elements.

The simplest way of implementing any form of pattern matching is to use function MATCH itself recursively. Consequently, if we come across the & sign in our pattern, we first attempt to match the rest of the pattern with the data as a whole (assuming that the & sign corresponds to the empty sequence) and, if that does not work, we attempt to match the rest of the pattern with each CDR in turn of the data list, until we have found a list that works. Here is the new code for our first version of MATCH:

```
(DEFUN MATCH (PATTERN DATA) (COND
            ((ATOM PATTERN)(EQUAL PATTERN DATA))
            ((EQUAL (CAR PATTERN) '&) (COND
                ((NULL (CDR PATTERN)) T)
                ((MATCH (CDR PATTERN) DATA))
                ((DATA (MATCH PATTERN (CDR DATA))))))
            ((ATOM DATA) NIL)
            ((EQUAL (CAR PATERN) '$)
                (MATCH (CDR PATTERN) (CDR DATA)))
            ((MATCH (CAR PATTERN) (CAR DATA))
                (MATCH (CDR PATTERN) (CDR DATA)))))
```

Note that we introduced the new clause *before* testing if the data is an atom. This is necessary whenever the &-sign matches an empty sequence at the end of the data.

Here are a few examples of calls with their results:

```
(MATCH '(A B C) '(A B C))           → T
(MATCH '(A $ C) '(A B C))           → T
(MATCH '(A $ C) '(A 1 C))           → T
(MATCH '(A $ C) '(A 1 2 3 C))       → NIL
(MATCH '(A & C) '(A 1 2 3 C))       → T
(MATCH '(A & C) '(A C))             → T
```

Look carefully at the difference in effect between $ and &: one matches *a single* element and the other matches *a sequence of elements* (none or more).

A very common technique in artificial intelligence programming is to try out calls to a function. Let us therefore look at an execution trace of the call

```
(MATCH '(A & C) '(A 1 2 3 C))
```

```
→ (MATCH (A & C) (A 1 2 3 C))  ;initial call
  → (MATCH A A)                ;first recursive call
  ← MATCH T                    ;its working
  → (MATCH (& C)(1 2 3 C))     ;second recursive call
    → (MATCH (C) (1 2 3 C))    ;try the empty sequence
      → (MATCH C 1)            ;is it working?
      ← MATCH NIL              ;no
    ← MATCH NIL                ;not the empty sequence
    → (MATCH (& C)( 2 3 C))    ;try a one-element sequence
      → (MATCH (C) (2 3 C))    ;and try again
```

```
    → (MATCH C 2)                ;not very promising;
    ← MATCH NIL                  ;as we were saying;
  ← MATCH NIL                    ;not a one-element sequence
  → (MATCH (& C) (3 C))          ;try a two-element sequence
    → (MATCH (C) (3 C)
      → (MATCH C 3)
      ← MATCH NIL                ;fail;
    ← MATCH NIL                  ;again
    → (MATCH (& C)(C))           ;what about three elements?;
      → (MATCH (C) (C))          ;so far so good;
        → (MATCH C C)            ;this time it's working;
        ← MATCH T                ;there we go
        → (MATCH NIL NIL)        ;nearly there;
      ←— MATCH T                 ;it really was working;
←———— MATCH T                    ;at last;
= T                              ;and here's the final result
```

We shall return to pattern matching in Chapter 18, where we shall set out to make it more general and powerful. But before we do that, we badly need to take another look at lists. That is what we shall be doing in the next chapter.

14.2 EXERCISES

1. Which of the recursive calls in MATCH are terminal recursive calls and which are not?

2. How could we modify program MATCH above to allow the use of patterns giving a choice between the various possible elements? For example, if we wanted to find all calls to functions CAR and CDR within a program we could represent those calls using the two patterns:

   ```
   (CAR &)
   (CDR &)
   ```

   What modification do we have to make to allow us to use a single pattern, e.g.:

   ```
   ((% CAR CDR) &)
   ```

   i.e.: each time our function finds that a pattern contains a sub-list beginning with the sign %, it knows that the rest of the list is a set of elements *one* of which must be in the corresponding position of the list given as data.

   Here are a few examples of the revised MATCH function with this new type of pattern:

(MATCH '(A (% B 1) C) *something*)

which only matches the lists (A B C) and (A 1 C);

(MATCH '((% A B)(% A B)) *something*)

which only matches the four lists: (A A) (A B) (B A) (B B).

15 ANOTHER LOOK AT LISTS

15.1 MEMORY ORGANIZATION IN LISP

How does LISP make use of computer memory? We know that LISP handles all sorts of objects: lists, numbers, character strings, atoms. Each type is held in a special memory zone on your computer.

Figure 1 shows how memory space is used in your LISP computer.[1] The numbers to the left of the table represent addresses. So, in our figure, the first α words of memory are used by the computer operating system. Naturally, this zone could be somewhere else; its precise location depends on the particular computer you are using. What matters is that *somewhere* there is a zone (or several zones) which cannot be used by LISP.

Let us now look at each of the LISP zones. The first, running from address α to address β-1, is occupied by the LISP interpreter itself. This is where the input/output programs, the little *top level* loop we met in Chapter 13 and so on, are held. This zone is not accessible to all users: it is the system zone of your LISP machine, just as the first zone is the system zone of your computer.

Next, there are zones for each of the types of objects used in LISP: a zone for storing numbers, another for character strings, another for atoms and a fourth for lists. Lists *only* appear in the list zone, numbers in the number zone, etc.

The final zone on your LISP machine is held for the stack. This is a work zone for the interpreter: it contains, for example, intermediate results. We shall see the importance of this zone later. For the moment we shall concentrate on the LIST zone, as it is the zone with which we shall be most concerned for the moment.

15.2 THE LIST ZONE

Lists were invented to use memory without size constraints. The basic idea is:

1. Organizing memory in a non-continuous way: a list does not necessarily have to occupy contiguous words in memory. For this to be possible, lists are implemented in such a way that, starting from *any* element in the list, it is always possible to find the next.

---

[1] The organization described here is not the only one possible; other LISP systems operate differently. We shall describe this organization because it is, conceptually, the easiest and nonetheless displays the most important characteristics of LISP memory organization.

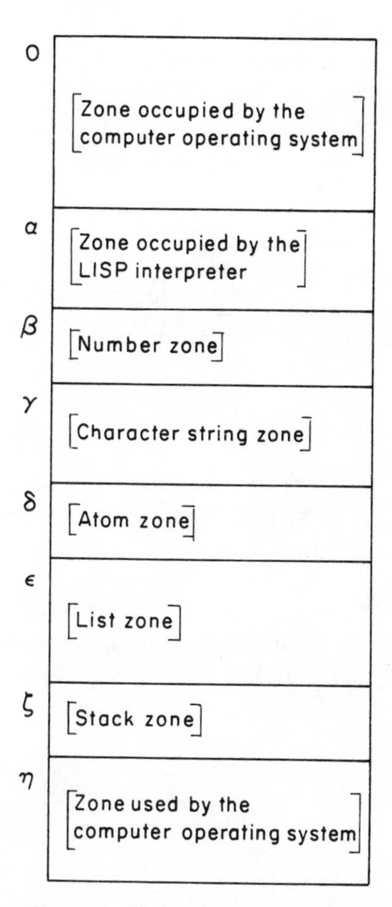

Figure 1: The various memory zones

2. Provide a *garbage collection* algorithm. This algorithm must be able to find lists (or parts of lists) which are no longer being used and include them once more in usable space.

All this may seem very abstract. Let us now look at the organization of this zone in more detail.

A list is made up of a CAR and a CDR, of a first element and the rest. These two parts reappear again and again within the machine, and therefore within the list zone; the CAR and CDR of a list occupy two adjacent words in the zone:

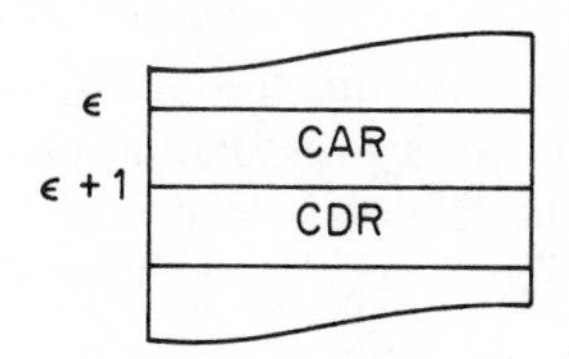

Each of these boxes represents a memory word in the list zone. The labels CAR and CDR within the boxes are *pointers* to the LISP objects forming the CAR and CDR of this list. Diagrammatically:

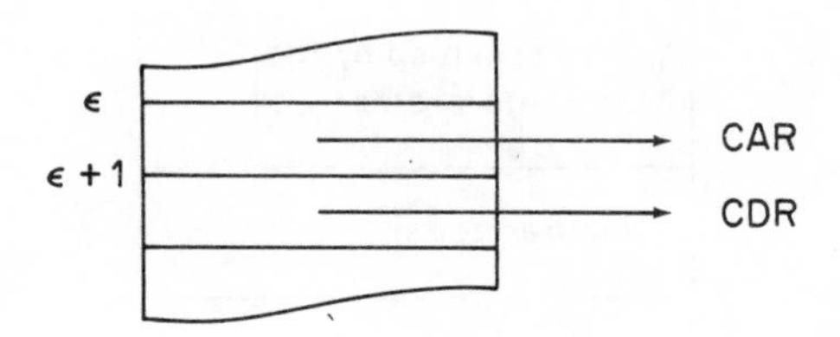

The CAR or CDR of a list may be of any type: atom or list. Consequently, CAR and CDR pointers can point to any LISP object *zone*. They are, therefore, in fact memory addresses.

Consequently, if we have a list (A B), formed of an atom A which is in memory at address δ and an atom B which is at address δ + 40, this list could be implemented in memory as follows:

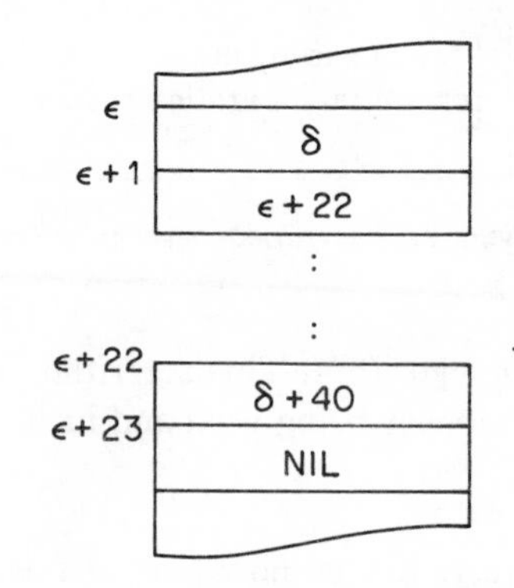

Lists are therefore ordered pairs of words made up of a pointer towards the location of the value of the CAR and a pointer to the location of the CDR, i.e. *the address* of the memory word in which the corresponding value is held. The end of a list is indicated by a pointer to the special atom NIL.

In order not to overload our diagrams with too many addresses, we shall represent pointers towards lists by arrows to the corresponding pair and the end of a list by a diagonal bar across the CDR of the final pair, and instead of giving the addresses of atoms, we shall simply show pointers towards the names of atoms. This, for the same list (A B), would give us:

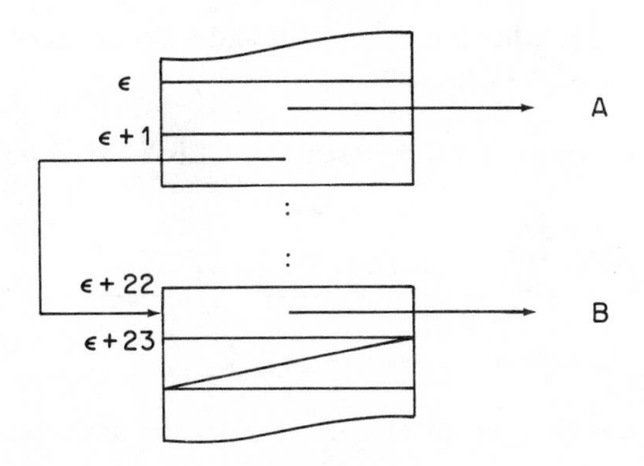

Clearly, we cannot constantly examine real addresses, which is why we shall use a diagrammatic representation of lists which only takes into consideration *ordered pairs* directly used in the list. A list will therefore be represented as a sequence of *ordered pairs* made up of the CARs and CDRs of that list:

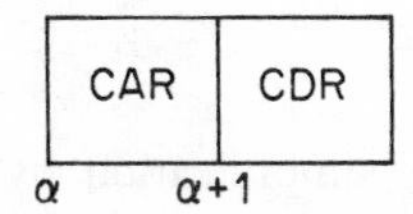

The list (A B) above would therefore be represented as follows:

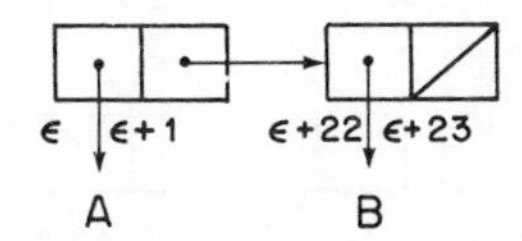

Normally, if we need to give a diagrammatic representation of a list, we shall ignore real implementation addresses and draw a list such as (A B) as follows:

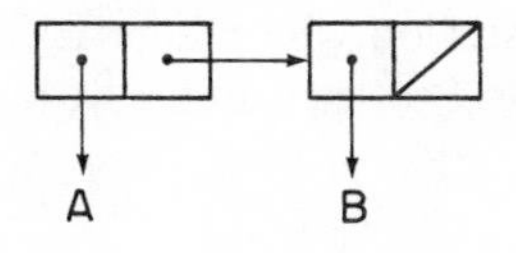

But, though we shall use this fairly abstract form of representation, it should not be forgotten that the internal representation in fact uses real addresses.

If this list (A B) is the value of variable L, instead of representing it diagrammatically as atom L with the address of this list as its C-value (cf. section 8.1), we shall simply use an arrow from L towards the list, as below:

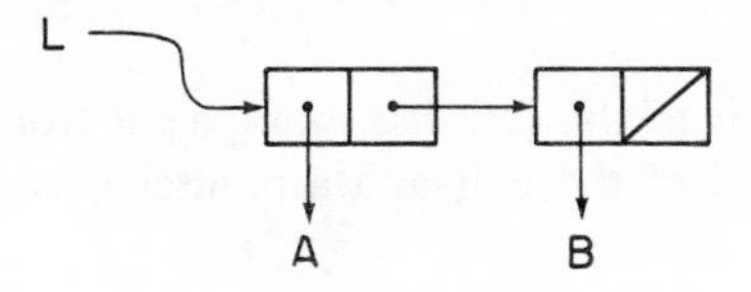

This makes it easy to see what functions CAR and CDR do: they return the pointers of the CAR and CDR fields of the list.

Let us look at a few examples of this representation. List (A B C) would be implemented as follows:

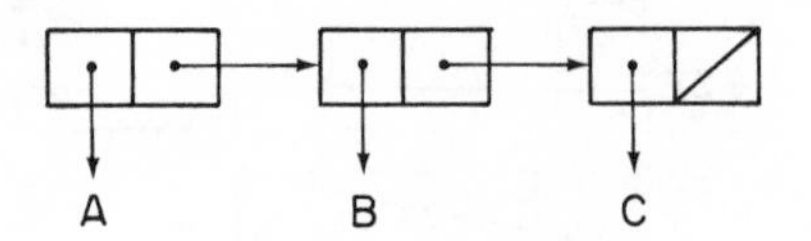

and list ((A B) C) would, in terms of ordered pairs, be implemented as follows:

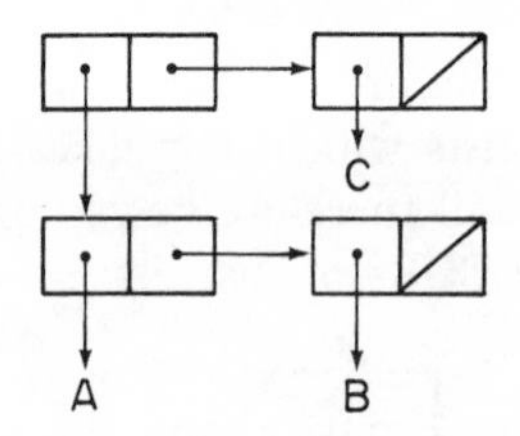

We shall now introduce a further simplification by only arrowing pointers towards lists. Our representation of the latter list therefore becomes:

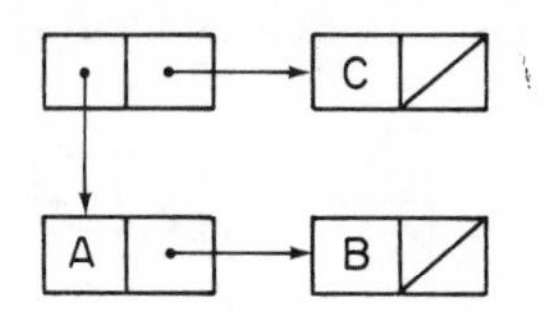

It should not be forgotten that every pointer and every atom is in fact simply an address, to the memory location where the corresponding object is held.

At the beginning of a LISP session, the whole LIST zone is organized as a *free list*, i.e. all ordered pairs are *chained* to each other to form an enormous list:

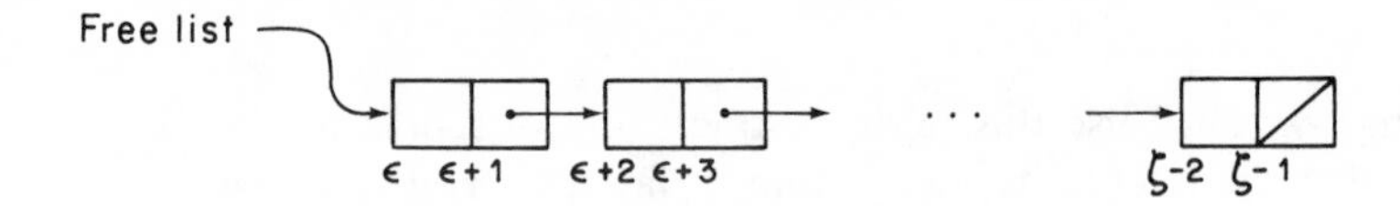

Every call to function CONS removes a pair from this free list. For example, if the first call to CONS is

(CONS 'A NIL)

CONS first looks for a pair in the free list, removes it from the list, inserts a pointer to atom A in the CAR field of the pair and a pointer to atom NIL in the CDR of the same field. This gives:

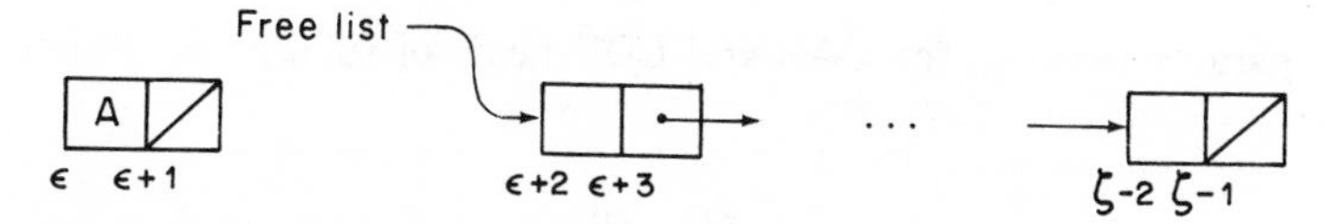

When the free list is empty, CONS sets a garbage collection mechanism going, producing a new free list from the pairs which are no longer needed, i.e. from pairs that are no longer being used.

Here are a further few examples of the representation of lists in the form of pairs. We start with the list ((A B) (C D)):

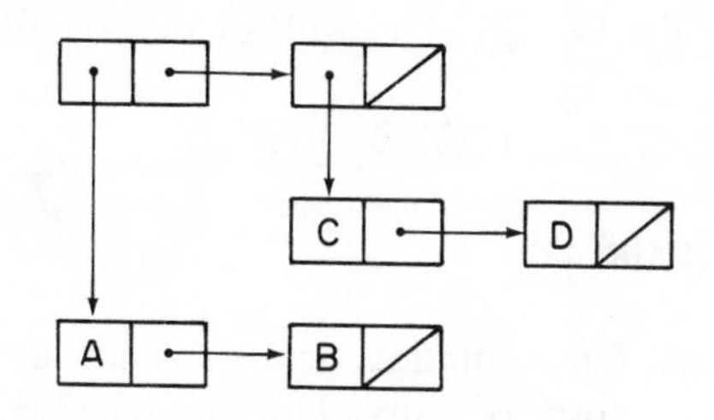

list ((A B (C D))):

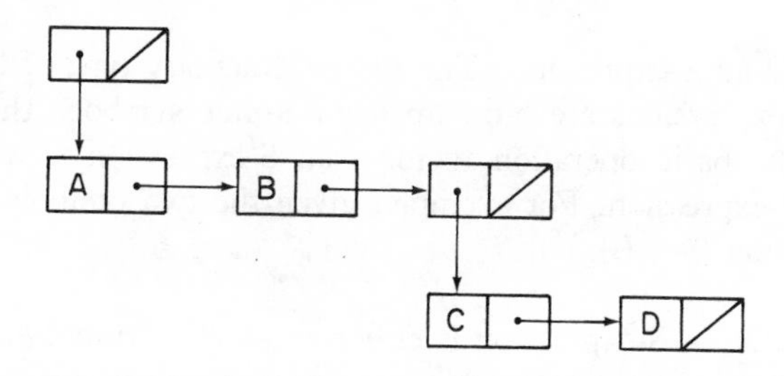

We said that the contents of the CAR or CDR field of a pair may be a pointer to any other LISP object. The CDR field of a pair may turn out to be a pointer not to a list but to an atom, as is the case for example in the pair below:

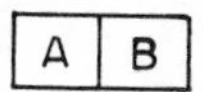

or in the list,

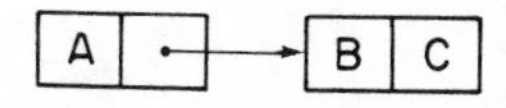

We do not yet have any way of representing these two lists externally. To read or write such lists, we have to introduce some further notation. To represent lists whose CDR field contains a pointer to an atom other than NIL[2] we shall use the character '.'

[2] Remember that a pointer to the atom NIL in the CDR field marks the end of a list.

(point) as separator between the CAR and CDR fields of such pairs. This gives us, for our first list,

(A . B)

and for our second list,

(A B . C)

Naturally, this changes our definition of function CONS in Chapter 2: there is no longer any reason why we should ask for the second argument of the CONS to be a list. On the contrary, the list (A . B) is a result of the following call:

(CONS 'A 'B)

and the CDR of this list is the *atom* B.

Clearly, we need an external representation, as a sequence of characters, to reflect the *implementation* of lists in the form of pairs. For this purpose, we shall return to the traditional definition of symbolic expressions (or S-expressions) used by John McCarthy, the inventor of LISP.

Any LISP object is an S-expression. The most elementary type of S-expression is an *atom*. All other S-expressions are built up from atomic symbols, the characters '(',')' and '.' (point). The basic operation to form an S-expression is to combine two to produce a larger S-expression. For example, given the two symbols A and B, we can form the S-expression (A . B).

The exact rule is that an S-expression is either an atomic symbol or a combination of the following symbols in the order given: an opening parenthesis, an S-expression, a point, an S-expression and a closing parenthesis.

Here, as examples of the application of this definition, are a few S-expressions and their internal representations in the form of ordered pairs:

A

clearly is an atom and therefore cannot be represented as an ordered pair.

(A . NIL)

(A . (B . NIL))

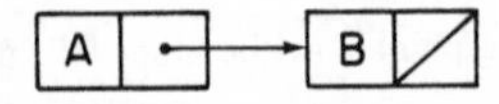

(A . B)

A | B

((A . B) . C)

· | C

A | B

((A . B) . (C . D))

· | · → C | D

A | B

((A . B) . (C . (D . E)))

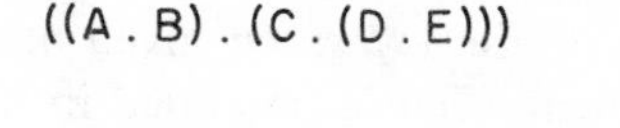

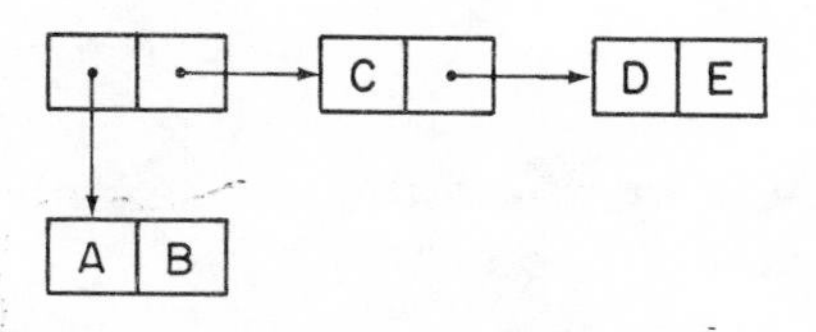

To move from this S-expression notation to the list notation to which we are used, all we have to do is:

1. Replace each occurrence of a point followed by an opening parenthesis by a blank and eliminate the corresponding closing parenthesis, which would transform, for example, (A . (B . NIL)) into (A B . NIL);

2. Eliminate occurrences of a point followed by NIL, which would for example transform (A B . NIL) into (A B).

Consequently, for the S-expressions above, we have the following correspondences:

A	≡	A
(A . NIL)	≡	(A)
(A. (B. NIL))	≡	(A B)
(A . B)	≡	(A . B)
((A . B) . C)	≡	((A . B) . C)
((A . B) . (C . D))	≡	((A . B) C . D)
((A . B) . (C . (D . E)))	≡	((A . B) C D . E)

Naturally, *pointer pairs*, i.e. pairs where the CDR is a pointer to an atom other than NIL, are still represented using the '.' character.

15.3 LIST MODIFICATION FUNCTIONS

The list processing functions we have met so far – CAR, CDR, REVERSE etc – do not modify the lists given as argument. Consequently, if at some point the value of variable L is the list (A B C), calculating the CDR of L will give us the list (B C), but the value of L will still be the old list (A B C).

In this section we shall study a number of *destructive* list processing functions, i.e. functions whose evaluation modifies the argument lists.

The two most important functions of this kind are RPLACA and RPLACD. They respectively *physically modify* the CAR or CDR of their first argument. Here are their definitions:

(RPLACA *list expression*) → *new list*
RPLACA physically replaces the CAR of the *list* given as first argument by the value of the *expression* given as second argument. The *list* modified in this way is returned as value.

(RPLACD *list expression*) → *new list*
RPLACD replaces physically the CDR of the *list* given as first argument by the value of the *expression* given as second argument. The *list* modified in this way is returned as value.

Here are a few examples:

```
(RPLACA '(A B C) 1))              →  (1 A B)
(RPLACA '((A B)(C . D)) '(1 2 3)) →  ((1 2 3) (C . D))
(RPLACD ' (A B C) 1)              →  (A . 1)
(RPLACD '((A B)(C . D)) '(1 2 3)) →  ((A B) 1 2 3)
```

The examples above use these functions to calculate a value. But it should be remembered that these two functions *modify* the lists given as argument. To make this clear, look at the following function:

```
(DE FOO (X)
    (CONS (RPLACA X 1) X))
```

What does this function do? Is the result the same if we replace the body by (CONS (CONS 1 (CDR X)) X)?

This needs to be examined in detail. If we call the function as follows:

```
(FOO '(A B C))
```

the variable X will be bound with the list (A B C). Before starting to evaluate the body of the function we therefore have:

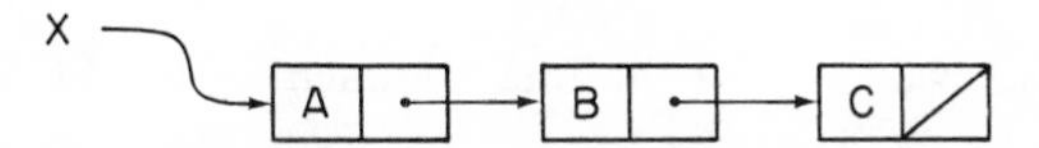

Evaluating the body of the function therefore means evaluating the CONS. We must therefore first evaluate the two arguments (RPLACA X 1) and X. RPLACA modifies list X, giving us:

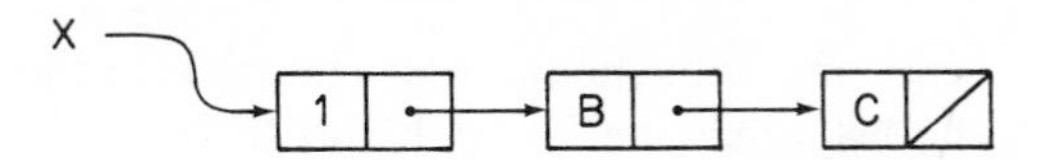

and it is this new list which will be the first argument for the CONS.

The evaluation of X (second argument) also produces the new list (1 B C), since that list is now the value of X. The result of a call to (FOO '(A B C)) is therefore the list:

```
((1 B C) 1 B C)
```

and not the list ((1 B C) A B C).

Be careful when you use these functions, since they are very dangerous: be sure you really want to *modify* the internal structure of the list.

Now suppose that the person who constructed program FOO had in fact wanted to construct a function producing a list whose CAR was the same list as the CDR with the first element replaced by atom 1, i.e. a function such that (FOO '(A B C)) returns ((1 A B) A B C). Seeing the error, the programmer tries a new version as below:

```
(DEFUN FOO (X)
           (LET ((X X)(Y X))
                      (CONS (RPLACA X 1) Y)))
```

Is this solution any better? The idea is obvious enough: since RPLACA modifies the list, we first save the value of X in variable Y, before applying RPLACA. Variable Y will then be the second argument of CONS. Let us see what happens, and call (FOO '(A B C)) again. As before, variable X will be associated with list (A B C):

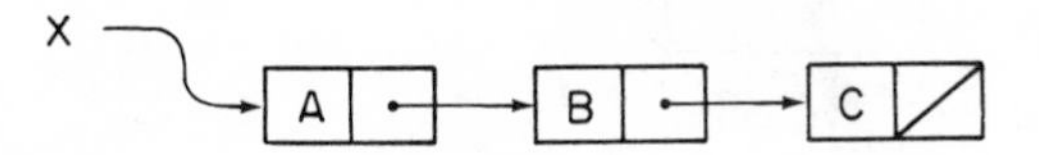

then, because of the LET function, X will be rebound with the same list and Y will be bound with the value of X, which gives us:

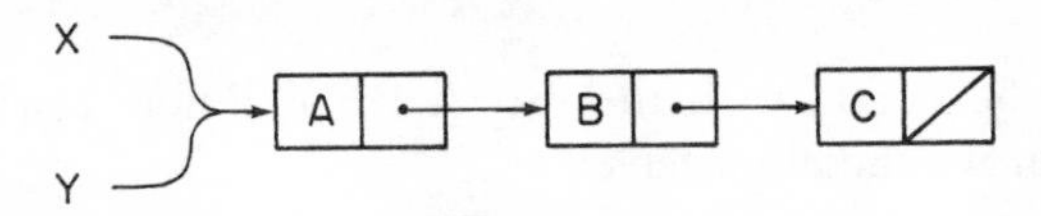

Then, as before, the evaluation of the first argument of the CONS gives list (1 B C) and modifies the value of X, giving us:

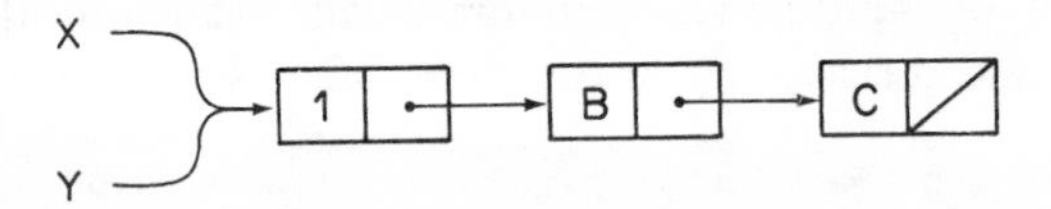

Since both X and Y are bound with the *same* list, i.e. the C-values of X and Y contain the same address in the list zone, the function RPLACA modifies the values of both variables. Consequently, even with this initial save of the value of X, nothing changes: the case is the same as before, i.e. the result will be the list ((1 B C) 1 B C).

As an exercise, we shall now write a function FREVERSE to invert a list *without* constructing a new list, i.e. without a single call to function CONS. To solve the problem, we change the pointers in the original list, in such a way that the same ordered pairs are chained as before, but with the chaining order reversed. The diagram below shows list L containing three elements A, B and C, with dotted lines showing how pointers should be after the list has been inverted; L′ is the value of L after inversion:

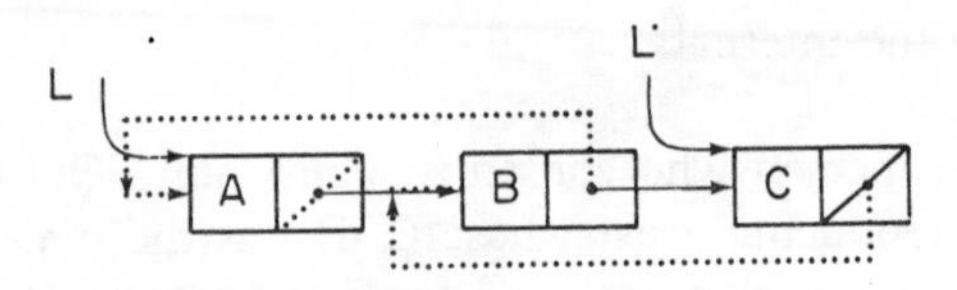

Let us take a look at a trace to see how we should proceed. At the start we have a pointer called L, to the initial list:

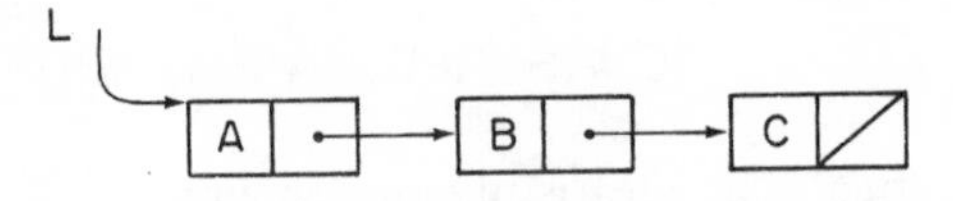

Clearly, the first thing to do is to replace the CDR of the first ordered pair by NIL, giving us:

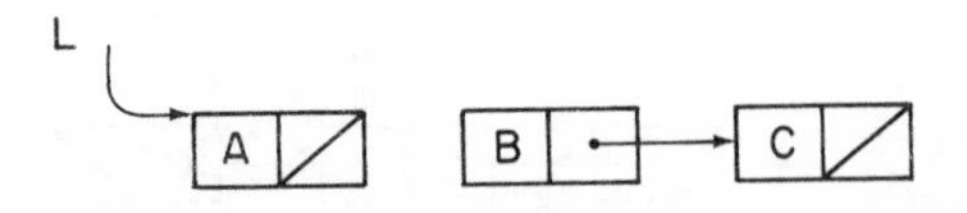

The next stage is to replace the pointer from the CDR of the second pair by a pointer to the first pair, which has just been modified. But after *replacing* the CDR of the first pair by NIL, we no longer have access to the second pair. Consequently, before carrying out this operation we have to save a pointer somewhere towards the rest of the list. Let us call this pointer NEXT. We therefore need the following configuration:

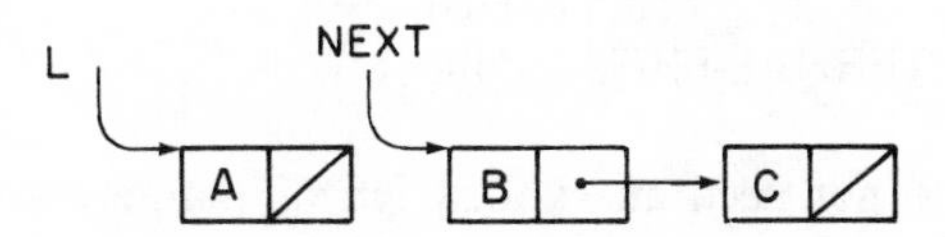

If we now replace the CDR of the second pair by a pointer to the value of L, we shall again lose the rest of the list. We must therefore start by moving the NEXT pointer on, but without losing its current value, since we shall later have to establish a pointer towards the pair pointed to at the moment by NEXT.

The solution is therefore to introduce a further pointer, which we shall call PREVIOUS. L can then point at any time to the new starting position of the list, while PREVIOUS points to the pair which, in the original list, preceded the element now pointed to by L. The pointer NEXT always indicates the remaining elements, which have not yet been modified. This gives us the following sequence of operations:

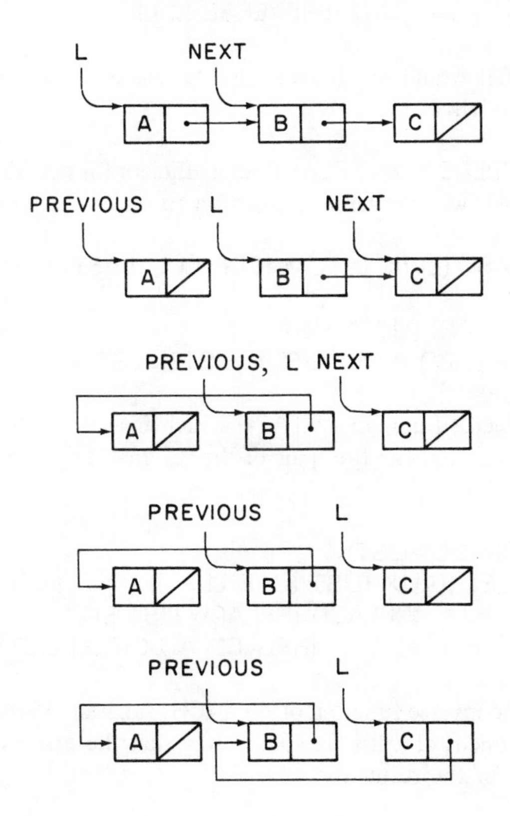

Here is the LISP definition of FREVERSE;

```
(DEFUN FREVERSE (L)
    (LABELS ((INTERNAL-FREV (L PREVIOUS)
                (IF L
                    (INTERNAL-FREV (CDR L) (RPLACD L PREVIOUS))
                    PREVIOUS)))
        (INTERNAL-FREV L NIL)))
```

In this version we do not need an explicit NEXT pointer, since the binding of L to the CDR of L, and therefore to the address of the following pair, takes place in parallel with the deletion of the CDR field itself.

Here then is a function which inverts a list without using a CONS, i.e. a function which needs no additional pair. FREVERSE will therefore be significantly quicker than function REVERSE considered in Chapter 6.

But it should never be forgotten that FREVERSE is *destructive*, and that the same care has to be taken as with function FOO above. This means that if, for example, we had the function:

```
(DEFUN BAR (L)
        (APPEND (FREVERSE L) L))
```

a call (BAR '(A B C)) would not longer give the list (C B A A B C), but the list (C B A A). Do you see why?

The two functions RPLACA and RPLACD are sufficient for any (destructive) operations on lists. For example, we can define a function to concatenate two lists:

```
(DEFUN LAST (L) (IF (ATOM (CDR L) L (LAST (CDR L))))

(DEFUN NCONC (LIST1 LIST2)
    (RPLACD (LAST LIST1) LIST2) LIST1)
```

We can now construct a function ATTACH which inserts a new element at the head of a list, in such a way that the first pair of the original list is also the first pair of the result list:

```
(DEFUN ATTACH (ELT LIST)
        (LET ((AUX (LIST (CAR LIST))))
            (RPLACD (RPLACA LIST ELT)
                (RPLACD AUX (CDR LIST)))))
```

We can also write the inverse function of ATTACH: function SMASH which physically removes the first element of a list, in such a way that the first pair of the original list is also the first pair of the result list:

```
(DEFUN SMASH (LIST)
            (RPLACD (RPLACA LIST (CADR LIST))(CDDR LIST)))
```

The common feature of all these functions is that they physically modify the structures of the list given as argument, and therefore the values of *all* variables pointing towards such a list are also modified.

15.4 FUNCTIONS EQ AND EQUAL

In Chapter 6 we built up function EQUAL, which tests the equality of two lists by scanning them in parallel and testing for equality between pairs of atoms using function EQ. This function therefore tests whether two lists have the same structure and whether they are made up of the same atoms in the same order.

The function EQ, we said, could only test for the equality of atoms. We ought to be more precise: LISP atoms exist in only one copy. Consequently, if we have the expression:

```
(CONS 'A '(A B C))
```

the two occurrences of atom A are two occurrences of a pointer to the *same* atom. To take another example, each occurrence of the atom CONS within a program refers to precisely the same atom. Consequently, the test of equality provided by function EQ is in fact a test of equality applied to two addresses (and not the test of equality for a character string). It is therefore a test of identity: the two atoms A in the expression above are references to one and the same object.

Knowing that EQ can be reduced to a comparison of addresses, and knowing that lists are also accessible by means of their addresses, we can use this function to test the *identity* of two lists.

To make this clearer, let us construct two lists called X and Y. The first, X, will be formed using the call

```
(LIST '(A B) '(A B))
```

and the second, Y, by the call:

```
(LET ((X '(A B)))(LIST X X))
```

Note carefully the difference between these two lists. Externally both are written:

```
((A B) (A B))
```

but internally the first is represented by:

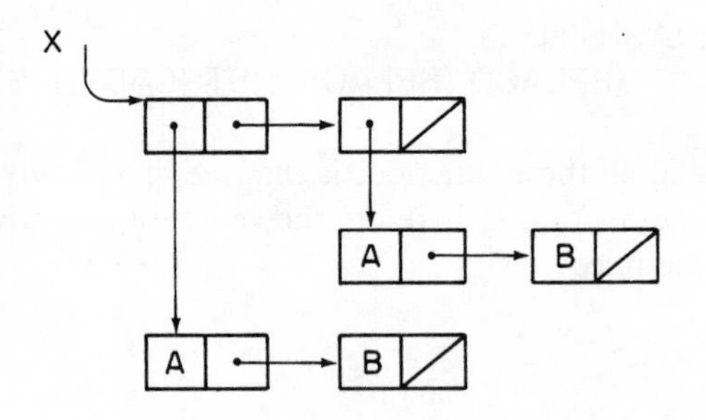

and the second by:

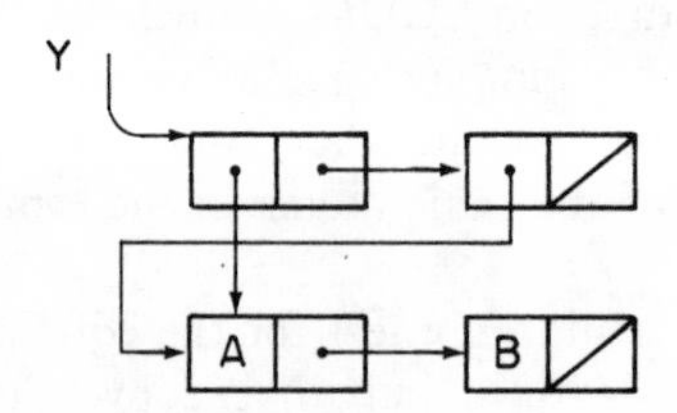

Though the two lists are very different, from the internal point of view, their externa written representation is identical.

The difference between list X and list Y becomes clear if we test for the identity o two sub-lists:

```
(EQUAL (CAR X) (CADR X))        →  T
(EQUAL (CAR Y) (CADR Y))        →  T
(EQ (CAR X) CADR X))            →  NIL
(EQ (CAR Y) (CADR Y))           →  T
```

Clearly, the first and second elements of Y are pointers to the same list: function EC therefore gives the answer T.

Such lists are said to be *shared*. Using the standard output options, there is no wa to distinguish between shared and unshared lists. But, of course, shared lists are alway produced by operations on lists (i.e. they cannot be entered directly from the terminal)

15.5 CIRCULAR LISTS

Let us look at a few more shared lists. What result will the following operatio produce?

```
(LET ((X '(A B C))) (NCONC X X))
```

NCONC concatenates two lists. This means that we are concatenating list X, i.e. (A B C), with list X itself, giving us:

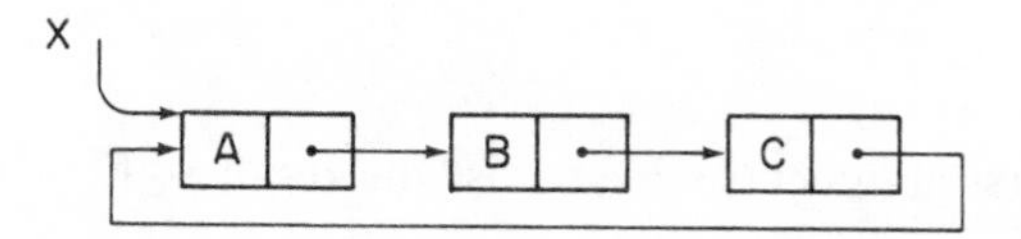

This is a *circular* list, an endless list. Printing it will give something like:

(A B C A B C A B C A B C . . .)

Consequently, it is impossible *a priori* to print such lists. Note that these lists can be very useful to represent infinite data structures, but they are very dangerous: many search functions run the risk of executing indefinitely on them. For example, if we were looking for atom D in this list, the call

(MEMQ 'D X)

would loop endlessly, since function MEMQ compares the first element of the list with the argument being searched for and tries again with the CDR of the list if there is no match, until it reaches either the element it is looking for or the end of the list. Since our list X has no end, a search for an element it does not in fact contain will never reach an end.

To write a function which searches for an occurrence of a given element in a circular list, we therefore have to maintain a permanent pointer to the start of the list, so that we can use function EQ to test whether the CDR of the list is EQual to the whole list, and therefore avoid infinite repetition of the search function.

The following is the definition of a predicate C-MEMQ testing for the occurrence of an element within a circular list:

```
(DEFUN C-MEMQ (OBJ L)
    (LABELS ((INTERN-C-MEMQ (LL OBJ L) (COND
                ((ATOM LL)(EQ LL OBJ))
                ((EQ (CAR LL) OBJ) T)
                ((EQ L (CDR LL)) NIL)
                (T (OR (INTERN-C-MEMQ (CAR LL) OBJ L)
                       (INTERN-C-MEMQ (CDR LL) OBJ L))))))
        (INTERN-C-MEMQ L OBJ L)))
```

If this function is called with atom D as first argument, and as second argument the circular list:

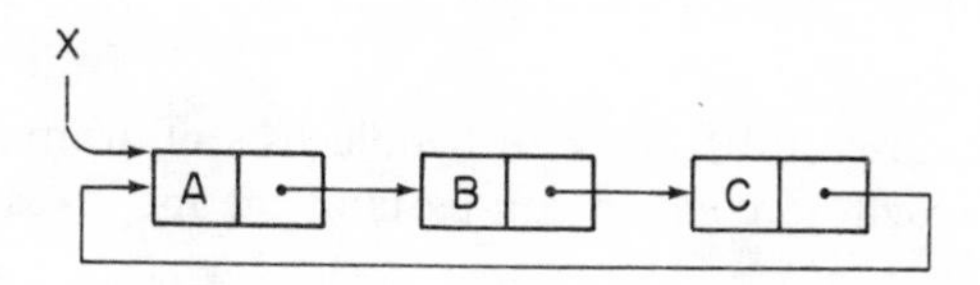

the result will be NIL.

Of course, if the first argument is in the list the result is T.

Let us now modify our list by means of the following operation:

```
(RPLACA (CDR X) X)
```

This operation gives us a doubly circular list: a loop on the CAR and a loop on the CDR.

The list is as follows:

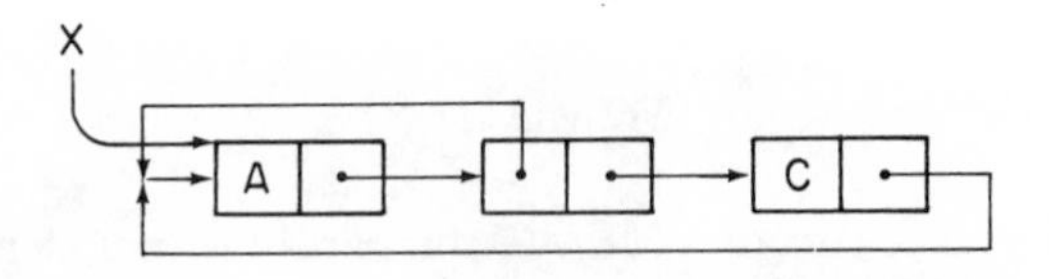

C-MEMQ, the search function for an element within a circular list, will no longer work on this list, as it does not test for the circularity of the CAR of the list.

To generalize the C-MEMQ function, we have to test both the CAR and the CDR of the list to check for circularity towards the start of the list. The new function is in Le_LISP:

```
(DE C-MEMQ (OBJ L)
        (IF (EQ (CAR L) OBJ) T
           (LETN SELF ((LL (CDR L))) (COND
                 ((ATOM LL) (EQ LL OBJ))
                 ((EQ L LL) ())
                 (T (OR (SELF (CAR LL)) (SELF (CDR LL))))))))
```

and in Common-LISP:

```
(DEFUN C-MEMQ (OBJ L)
    (LABELS ((INTERN-C-MEMQ (LL OBJ L) (COND
                ((ATOM LL)(EQ LL OBJ))
                ((EQ L LL) NIL)
                (T (OR (INTERN-C-MEMQ (CAR LL) OBJ L)
                       (INTERN-C-MEMQ (CDR LL) OBJ L))))))
       (IF (EQ (CAR L) OBJ) T
          (INTERN-C-MEMQ (CDR L) OBJ L))))
```

Finally, we can generalize still further to handle lists of arbitrary circularity, i.e. to forms of circularity which do concern sub-parts of the list, as is the case for example in list X below:

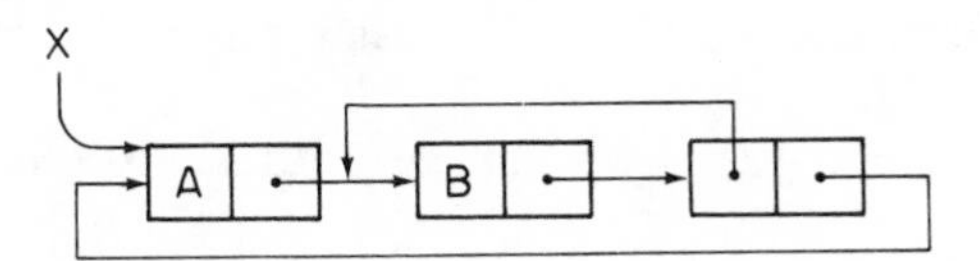

In these general cases, as well as overall circularity tests starting from CAR and CDR fields, a stack has to be maintained of all sub-lists met, in order to spot circularity on any substructure. This gives us:[3]

```
(DEFUN C-MEMQ (OBJ L &OPTIONAL AUX) (COND
                         ((ATOM L)(EQ L OBJ))
                         ((MEMQ L AUX) NIL)
                         (T (OR (C-MEMQ OBJ (CAR L) (CONS L AUX))
                                (C-MEMQ OBJ (CDR L) (CONS L AUX))))))
```

Clearly, function MEMQ used in the line

```
((MEMQ L AUX) ())
```

is the standard MEMQ function, whose definition is in exercise 4 of chapter 6: it is a function which tests for equality using function EQ.

15.6 ASSIGNMENT FUNCTIONS

After so many modification functions, we shall end this chapter with a brief look at functions that modify the values of variables.

So far, the only way that we know to assign values to variables is *dynamic binding*, i.e. linkage of values to variables during calls to user functions. In dynamic binding, when a function is called, variables take new values, determined by the arguments of the call, and when the function is exited, the old values of these variables are restored.

In LISP, there are two other ways of assigning values to variables, using the *assignment* functions SETQ and SET. Function SETQ takes at least two arguments, the first of which is the name of the variable, the second any LISP expression.

The *effect* of the evaluation of a call to function SETQ is to evaluate the expression passed as second argument, and assign the value returned as the new value of the variable passed as first argument. It is therefore a function modifying the C-value of atoms, with no possibility of automatic restoration later. The *value* of the evaluation of a call to function SETQ is the result of the evaluation of the expression.

[3] This version of C-MEMQ is based on Patrick Greussay's function SKE.

Here is an example of interaction with LISP using this function:

```
? (SETQ X 1)                    ;X takes the numeric value 1
=1
? X                             ;find out the value of X
=1
? (SETQ X (+ X 3))
=4
? X                             ;find out the new value of X
=4
```

and here is an elegant way of exchanging the numeric values of X and Y without using an intermediate variable:

```
? (SETQ X 10)                   ;initial value of X
=10
? (SETQ Y 100)                  ;initial value of Y
=100
? (SETQ X (- X Y))
=-90
? (SETQ Y (+ X Y))
=10
? (SETQ X (- Y X))
=100
? X                             ;check that the value of X
=100                            ;really is the old value of Y
? Y                             ;and that the value of Y
=10                             ;really is the old value of X
```

Function SETQ is principally used for the construction and modification of databases used by a whole series of functions, as it allows assigning values to variables from outside any of those functions, i.e. to global variables.

Note finally that in general function SETQ will take 2*n arguments. Here is the formal definition of SETQ:

(SETQ var_1 val_1 . . . var_n val_n) → *value* (val_n) and the C-value of var_1 is the result of the evaluation of val_1. . ., the C-value of var_n is the result of the evaluation of val_n.

Function SET is identical to SETQ except that it evaluates its first argument, an evaluation which returns the name of a variable. By way of definition of function SET we can say that the expressions

(SETQ *variable expression*)

and

(SET (QUOTE *variable) expression*)

are equivalent. Function SET allows *indirection*. For example, the sequence of instructions:

```
(SETQ A 'B)
(SET A 3)
```

will give variable B the numeric value 3.

To provide an example, here is a further version of function REVERSE, distinguished by the fact that it modifies the value of the variable given as argument. This is a Le_LISP function, we will see in the next chapter, how to do these kinds of things in Common-LISP.

```
(DF PREVERSE (L RES)
     (LETN SELF ((LST (EVAL (CAR L))))
                    (IF (NULL LST) (SET (CAR L) RES)
                         (SETQ RES (CONS (CAR LST) RES)
                              LST (CDR LST))
                         (SELF LST))))
```

Here is an example of the use of this function:

```
? (SETQ K '(A B C))        ;to assign a value to variable K
= (A B C)
? (REVERSE K)              ;call the standard REVERSE function
= (C B A)                  ;result
? K                        ;look at the value of K
= (A B C)                  ;as expected
? (PREVERSE K)             ;call the new function
= (C B A)                  ;result
? K                        ;look at the value of K
= (C B A)                  ;the value has indeed been modified
                           ;which is the difference between the
                           ;functions
```

Moreover, function SETQ can be defined in Le_LISP using function SET as follows:

```
(DF SETQ X
  (LET((VARIABLE (CAR X)) (VALUE (EVAL (CADR X))) (REST (CDDR X)))
       (SET VARIABLE VALUE)
       (IF REST (EVAL (CONS 'SETQ REST)) VALUE)))
```

To finish this chapter on impure LISP programming, let us have a short look at Common-LISP's general assignment function: the function SETF. We have seen this function already in Chapter 8, as a means to define the function PUTPROP. Here we will examine it more generally.

First let us just review the various functions for changing LISP data structures we have seen already. They are SET and SETQ, the two we have just examined and RPLACA and RPLACD, the functions to change the CAR or CDR of a list. All these functions may be re-written using the SETF function. For example:

```
(SETQ X 1)
```

does exactly the same as:

```
(SETF X 1)
```

and

```
(SET 'X 1)
```

does exactly the same as:

```
(SETF (SYMBOL-VALUE 'X) 1)
```

knowing that the function SYMBOL-VALUE returns the value of a symbol.

SETF expects two arguments: a *accessor* function, such as SYMBOL-VALUE in the example above, and the new *value* for the accessed part. The effect of a call to SETF is to change the value of the accessed part of the structure. SETF functions with almost all standard accessing functions. Here are some more examples of calls to SETF and their equivalent expressions using standard functions we already know:

```
(SETF (CAR X) 1)
```

this is the same as changing the first element of the list X to 1, that is, it is the same as:

```
(LET ((G824 X) (G825 1)) (PROGN (RPLACA G824 G825) G825))
```

which is just a complicated way of saying (RPLACA X 1).

Here is a RPLACD equivalent:

```
(SETF (CDR X) '(A B C))
```

which could translate in something like:

```
(LET ((G833 X) (G834 '(A B C)))
          (PROGN (RPLACD G833 G834) G834))
```

Once more, a very complicated way to express the same as (RPLACD X '(A B C)).

Until now, all the applications of SETF we have seen, have been instances of standard function calls. But SETF is more general: it accepts all *access* functions and assigns values to the corresponding parts of LISP structures. We have seen the PUTPROP equivalent with

```
(SETF (GET X Y) value)
```

Here is another example:

```
(SETF (CDDR X) '(A B C))
```

does the same as:

```
LET (G824 X) (G825 '(A B C)))
        (PROGN (RPLACD (CDR G824) G825) G825))
```

which, once more, is just a complicated way to write (RPLACD (CDR X) '(A B C)).

What happens actually, is that Common-LISP translates calls to the SETF function into calls to standard functions. So all the example calls of SETF we gave above, are effectively translated into their equivalent expressions. These kind of functions, functions which translate their calls first into other expressions, are called *macro-functions*. Macro-functions are the subject of the next chapter. Here, let us just note that SETF accepts as accessing function all standard accessing functions, such as GET, CAR, CDR, CADR, etc, as well as user defined accessing functions, if these user defined functions are defined as macro-functions.

If you don't remember the exact function changing values of some structure, or some sub-part of a structure, you can always replace it by a call to the general assignment function SETF.

15.7 EXERCISES

1. Show graphically what the effect of the following instructions is:

   a. (SET 'X '(A B C)) (ATTACH 1 X)

   b. (LET ((X '(A B C))) (SMASH X))

   c. (SETQ X '(A B C)) (RPLACA (RPLACD X X) X)

2. Write function FDELQ with two arguments ELT and LST, which physically removes all occurrences of element ELT inside list LIST (be careful with the case where the list begins with element ELT).

3. Write a function which will physically invert a simple circular list.

4. Modify function FREVERSE so that it will invert a list at all levels.

Exercises for Le_LISP only:

5. The sequence of instructions

   (SETQ *variable* (CAR *a-list*))
   (SETQ *a-list* (CDR *a-list*))

   is very common. This is why many LISP systems have a function NEXTL defined as follows:

   (NEXTL *variable*) → (CAR (*value* (*variable*))) and *variable* takes as new value (CDR *variable*).

   With this function, function PREVERSE may be redefined as follows:

```
(DF PREVERSE (L RES)
    (LETN SELF ((LST (EVAL (CAR L))))
                (IF (NULL LST) (SET (CAR L) RES)
                               (SETQ RES (CONS (NEXTL LST) RES))
                               (SELF LST)))))
```

   Define function NEXTL.

6. Define function RPLACB which combines functions RPLACA and RPLACD, replacing the CAR of the list given as first argument by the CAR of the list given as second argument, and the CDR of the list given as first argument by the CDR of the list given as second argument. For example, if variable X is assigned the value (X Y Z), the result of the call

   (RPLACB X '(1 2 3))

   assigns the value (1 2 3) to variable X. Furthermore, the first pair in the original list is at the same address as the first pair in the result list.

   In LE_LISP, this function is called DISPLACE.

16 MACRO-FUNCTIONS

During the construction of large programs we often find that we have to construct relatively simple functions just to increase the readability of programs. So, if in the course of your program you often have to find the fourth element of a list, rather than constantly writing:

(CADR (CDDR *a-list*))

you will probably define a function called 4-TH, defined as follows:

(DEFUN 4-TH (L) (CADR (CDDR L)))

and replace all occurrences of the instructions above by:

(4-TH *a-list*)

to make the program more readable.

The problem with this way of programming is that the execution of the program becomes slower and slower: each call to a user function initiates the save, binding and restoration process for the values of the variables. Because of this dichotomy between readability and execution time, LISP supplements functions DE (and, in Le_LISP, DF) with another type of function: these are *macro-functions* or simply *macros*. This chapter presents one aspect of their use.

Macro functions are defined using the function DEFMACRO, for *DEF*ine *MACRO* function. These functions have the same syntactic form as all other definition functions: a name, a list of variables and a body:

(DEFMACRO *name list-of-variables body*) → *name*
defines a function of the macro type

The characteristic property of macro-functions is:

1. When called, the variables are bound to the corresponding arguments, *without evaluating* them. In Common-LISP, where we don't have the DF function defining form (see Chapter 11.1.2), this is the only means to define functions which do not evaluate its arguments.

2. The body is evaluated and the result of this evaluation is *again* evaluated. The evaluation of a call to a macro-function therefore occurs in two stages.

Clearly, the very originality of macro-functions is the double evaluation mechanism. This means that the first evaluation of a macro-function call should return a LISP form, i.e. a piece of code.

Let us start with a very simple example. Here is the macro-function EXAMPLE1 which, whenever it's called returns the value of (1+ 2):

```
(DEFMACRO EXAMPLE1 () '(1+ 2))
```

This macro simply returns the value of the expression

```
(QUOTE (1+ 2))
```

that is the list (1+ 2) in the first stage of its evaluation. That list is then re-evaluated in the second stage, giving us the result 3 for any call to EXAMPLE1.

But let us return to our function 4-TH and define a macro-function to meet our needs. Here is the first version:

```
(DEFMACRO 4-TH (L) (LIST 'CADR (LIST 'CDDR L)))
```

Note that the body of this macro constructs an expression to evaluate. That evaluation gives another expression, which computes the CADR of the CDDR of the argument given. Clearly, we could have written the body of the function more simply using the macro-character *backquote*, giving us:

```
(DEFMACRO 4-TH (L) `(CADR (CDDR ,L)))
```

If we call this macro as follows:

```
(4-TH '(A B C D E F))
```

the result of the first evaluation of this call is the expression:

```
(CADR (CDDR (QUOTE (A B C D E F))))
```

and the next evaluation of this expression returns atom D, which is the result of this call to our macro 4-TH.

We can, of course, also construct recursive macro-functions. For example, the macro MCONS below calls itself recursively until it reaches the end of the list passed as an argument:

```
(DEFMACRO MCONS L
    (IF (NULL (CDR L)) (CAR L)
        '(CONS ,(CAR L) (MCONS ,@(CDR L)))))
```

Note the degenerated *list of variables* in this macro definition. It is not a list at all, it is just a variable. We have already met similar lists of variables in Le_LISP. In Common-LISP, this is the only function defining form allowing *destructuring* variable lists. This kind of variable list is called *destructuring*, because the binding of the variables is guided by the structure of the variable list.

If, as is the case here, the list of variables is reduced to a single variable, then the list of all the arguments to the macro is bound to this variable. For example, if we could call MCONS like:

```
(MCONS A B C D E F G)
```

then the variable L would be bound to the list:

```
(A B C D E F G)
```

Since MCONS is a macro, none of the arguments is evaluated. Naturally, we could have defined MCONS as well like:

```
(DEFMACRO MCONS (&REST L) . . .)
```

Below we will study some more example of *destructuring* lists of variables and the possible binding. Consider a macro definition like:

```
(DEFMACRO FOO ((X Y Z) ((FTN1 ARG1)(FTN2 ARG2) . REST) THE-LAST) . . .
```

Now consider this macro call:

```
(FOO (1 2 3) (((4 VIER FOUR QUATRE) 5)(6 7) 8 9 10) 11)
```

This would cause the following bindings:

variable	*value*
X	1
Y	2
Z	3
FTN1	(4 VIER FOUR QUATRE)
ARG1	5
FTN2	6
ARG2	7
REST	(8 9 10)
THE-LAST	11

Here again, the dotted pair could have been expressed equivalently with the &REST specifier.

If we would call the FOO like:

```
(FOO (1 2) ((4 VIER FOUR QUATRE) SIX) 10)
```

LISP would give an error message since this call doesn't match the list of variables in several places:

1. There is no argument for the parameter Z
2. The arguments FOUR and QUATRE don't have corresponding parameters.
3. Finally, there is the atom SIX at a place where LISP expects a list. Even if the value of SIX would be a list of two elements, this wouldn't work, since LISP expects that the structure of the macro call itself matches that of the DEFMACRO parameter list. Often the list of parameters, or of variables is called a *lambda-list*.

Note that in Le_LISP all *lambda-lists* are destructuring, in Common-LISP only the *lambda-lists* of DEFMACRO are destructuring.

Let us come back to our function MCONS. Here it is again:

```
(DEFMACRO MCONS L
    (IF (NULL (CDR L)) (CAR L)
        `(CONS ,(CAR L) (MCONS ,@(CDR L)))))
```

Note the form of the recursive call. If we would have written:

```
(MCONS ,(CDR L))
```

we would have constructed a call where MCONS receives but *one* argument, the list containing all its arguments except the first. This would not give the correct result. So, what we did is to construct a call to MCONS in such a way that it has just one argument less than initially.

Here is a trace of a call to MCONS:

```
? MCONS 1 2 3 4 5 NIL)                          ;the initial call
  → MCONS (1 2 3 4 5 NIL)                       ; the trace of the initial call
  ← MCONS (CONS 1 (MCONS 2 3 4 5 NIL))          ; its result
  → MCONS (2 3 4 5 NIL)                         ; the first recursive call
  ← MCONS (CONS 2 (MCONS 3 4 5 NIL))            ; its result
  → MCONS (3 4 5 NIL)
  ← MCONS (CONS 3 (MCONS 4 5 NIL))
  → MCONS (4 5 NIL)
  ← MCONS (CONS 4 (MCONS 5 NIL))
  → MCONS (5 NIL)
```

```
    ← MCONS (CONS 5 (MCONS NIL))
    → MCONS (NIL)
    ← MCONS NIL
 = (1 2 3 4 5)
```

Each time the function is called, the variable is bound to the list of the *unevaluated* arguments and evaluation takes place in two stages. Each recursive call is carried out *after* exiting the previous call. This is not really a recursive execution!

When we introduced macro functions, we justified their use on the basis of savings in execution time. With what we have seen until now, it cannot, however, really be claimed that they lead to any improvement. Far from it, macro-functions execute more slowly still than normal user functions.

What we did not mention until now is that macro-functions modify the functions they are called in: *the result of the first evaluation replaces the call to the macro-function.* Consequently, calls to macro-functions delete themselves, ensuring in this way that the calls are eliminated.

Let us return to our macro 4-TH:

```
(DEFMACRO 4-TH (L) '(CADR (CDDR ,L)))
```

To see clearly what *effect* this macro has, we shall construct a function FOO which calculates the sum of the first and fourth argument. This presents no difficulty:

```
(DEFUN FOO (&REST L) (+ (CAR L) (4-TH L)))
```

Here is an example of a call to this function:

```
(FOO 1 2 3 4 5 6 7)
```

and the result will, of course, be the numeric value 5. If we look now at the function FOO, we find it has now the following form:

```
(DEFUN FOO (&REST L)
       (+ (CAR L) (CADR (CDDR L))))
```

All further calls to function FOO will no longer involve a call to the macro 4-TH, since the first effect was to replace the initial call itself by the result of the evaluation of the body of the macro-function.

We can now produce a macro for the function NEXTL, introduced in the exercises for the last chapter:

```
(DEFMACRO NEXTL (VAR)
     (LET ((VAL (GENSYM)))
          `(LET ((,VAL ,VAR))
                (SETQ ,VAR (CDR ,VAL)) (CAR ,VAL))))
```

This definition needs some explanation. First, we use here a new function, the function GENSYM. This is a function which GENerates a new SYMbol. We use it every time we need a symbol which we are sure won't be used somewhere else. Here are some examples of calls to gensym:

```
?   (GENSYM)                      ; a call
=   G117                          ; a first symbol
?   (GENSYM)                      ; a another call
=   G118                          ; a different symbol
?   (GENSYM)                      ; yet another call
=   G119                          ; yet another symbol
```

Definitely, this is a very useful function for generating new, until now non-existing, symbols. We use the function GENSYM here, since we want to create a local variable VAL, which we want to be sure doesn't interfere with whatever variables are already existing.

The first evaluation of a call to NEXTL will generate a little program which initializes the newly generated variable to the value of the variable which is the argument to NEXTL, then assigns a new value to the variable given as an argument and, finally, returns the first element of the list bound to the *gensymed* variable.

Let us look at some simple calls and their effect:

```
?   (SETQ X '(A B C D))           ; let's give a value to X
=   (A B C D)
?   (NEXTL X)                     ; the first call
=   A                             ; the result
?   X                             ; what's the value of X?
=   (B C D)                       ; good, it changed
?   (NEXTL X)                     ; another call
=   B
?   X
=   (C D)                         ; again X changed its value
```

The function MACROEXPAND takes as argument an expression which may be a call to a macro. If it is, MACROEXPAND will apply the macro to its argument to produce the result of the first evaluation of the corresponding macro call. That is, MACROEXPAND shows what the code looks like once the macro has been *expanded*. Let's look at a possible expansion of a call to NEXTL:

```
(MACROEXPAND '(NEXTL X))
```

will return:

```
(LET ((G826 X)) (SETQ X (CDR G826)) (CAR G826))
```

Each call to NEXTL will produce a piece of code similar to the one above.

Macro-functions are essential if you want to write readable programs without losing efficiency. The first execution of a function containing a macro will be a little slower than the execution of a normal function. On the other hand, later executions will all be considerably faster.

16.1 EXERCISES

1. Define the macro-functions INCR and DECR which, like macro-function NEXTL, take the name of a variable as argument and return as result the value of that variable plus or minus 1. These two macros have the side effect of incrementing or decrementing the value of the variable given as argument.

2. Define the macro-function SETNTH as specified by the following examples:

```
(SETNTH '(A B C D) 3 'G)        →  (A B G D)
(SETNTH '(1 2 3 4) 4 5)         →  (1 2 3 5)
```

   and if variable X is bound to the value (1 2 3 4), after the call:

```
(SETNTH X 1 'A)
```

   the value of X will be the list (A 2 3 4).

17 VARIOUS FORMS OF REPETITION

So far, the only method we have met for carrying out repetitive operations, is recursion, i.e. an explicit call to a procedure while that procedure is already executing. In this chapter, we shall examine other ways of repeating a sequence of instructions and of calling functions.

17.1 IMPLICIT FUNCTION CALLS (BRANCHING)

The first form of function call, specific to interpreted LISP, is to use the program representation structure as a control structure. As LISP programs are represented in the form of lists, we can use the possibilities provided by lists to express repetition. How? Simply by remembering that repetition means doing the same things several times, and that a list can contain the same thing several times, as we saw in Chapter 15.

Take as example a program which reverses a non-circular list:[1]

```
(SETQ E
      '(IF (NULL L) M
           (PROGN (SETQ M (CONS (NEXTL L) M)) *****)))
(RPLACA (LAST (CAR (LAST E))) E)
```

Here we have constructed a circular list E with the following form:

```
(IF (NULL L) M
   (PROGN (SETQ M (CONS (NEXTL L) M))
          (IF (NULL L) M (PROGN (SETQ M (CONS (NEXTL L) M))
                                (IF (NULL L) M (PROGN (SETQ M ....
```

It may be regarded as a program name, and an example of a call is:

```
(SETQ M NIL L '(A B C D))  ; to assign values to the variables
(EVAL E)                   ; to launch execution of the program
```

This gives the list (D C B A) as result and assigns that result to variable M.

[1] This example is inspired by Daniel Goossens.

Repetition, and therefore the calls to the function, simply takes place by continuous reading of a circular list.

The function PROGN we used in this example, is a standard LISP function with an arbitrary number of arguments, which evaluates one argument after the other, and returns the value of its last argument. This function is mainly used to syntactically combine several expressions into one. Since the IF function of Common-LISP admits only three arguments (a *test*, a *then-action* and a *else-action*), and since we have here more than one *else-action*, we used the function PROGN to combine the two else-actions into one.

The functions NEXTL and LAST are those we have defined in previous chapters.

Another example showing how a control structure may be expressed within the structure of the internal representation of programs, is the following version of REVERSE:[2]

```
(SETQ REVERSE
     '(IF (NULL L) ()
         (LET ((X (NEXTL L))) (NCONC1 (EVAL *) X))))
(SETQ * REVERSE)
```

A call to this REVERSE function takes place, as before, by initialization of variable L and a call to (EVAL REVERSE).

The difference from function E is that in E we had a circular list, while in REVERSE circularity is obtained by explicit evaluation of a variable pointing towards the structure itself: this is a kind of indirect circularity.

Naturally, except in certain types of macro-functions, we rarely have to use this method of constructing programs, only shown here as a matter of curiosity (and also to illustrate the power of program representation in the form of lists).

17.2. FUNCTIONAL ARGUMENTS

Apart from the two function calls that we have just seen, all those we have met so far are calls in which the *name* of the function appears explicitly, as, for example, in the call (FOO 1). In other words, in most programs that we have seen, simple static reading can show what functions have been called.

LISP also provides for computation of the functions we may want to call: we already have met one way of doing so through the functions EVAL and APPLY. In the rest of this chapter, we shall examine other ways of computing function calls.

[2] Function NCONC1 can be defined as:

```
(DEFMACRD NCONC1 (L ELE) '(NCONC ,L (LIST ,ELE)))
```

Suppose we want to write a function that sorts a list of numbers in increasing order
A possible way to do that is as follows:

```
(DEFUN SORT-A-LIST (L)
       (INSERT (CAR L) (SORT-A-LIST (CDR L))))

(DEFUN INSERT (ELE L) (COND
       ((NULL L) (LIST ELE))
       ((< ELE (CAR L)) (CONS ELE L))
       (T (CONS (CAR L) (INSERT ELE (CDR L))))))
```

Here are two example calls:

```
(SORT-A-LIST '(12 0 3 1 −5 6))   →  (−5 0 1 3 6 12)
(SORT-A-LIST '(1 2 3 3 2 1))     →  (1 1 2 2 3 3)
```

Now suppose we want to write a function that sorts a list of numbers in decreasing order. With what we have seen, this second function may use the same algorithm as the one just used. The only modification to it would be to change the predicate $<$ in the predicate $>$. This would give us the second function:

```
(DEFUN SORT-A-LIST-IN-DECREASING-ORDER (L)
       (INSERT-1 (CAR L)
       (SORT-A-LIST-IN-DECREASING-ORDER (CDR L))))

(DEFUN INSERT-1 (ELE L) (COND
       ((NULL L) (LIST ELE))
       ((> ELE (CAR L)) (CONS ELE L))
       (T (CONS (CAR L) (INSERT-1 ELE (CDR L))))))
```

If, in addition, we want to write a function which sorts a list of characters, we should still once more rewrite the function, using this time the predicate CHAR< or CHAR>, depending if we want to sort them in increasing or decreasing order.

This is rather awkward: since the algorithm used in all these different functions is the same, and since only one of the predicates changes, what we would like to do, is give this predicate as an argument to the function. In this way we could call SORT-A-LIST with two arguments, a list to sort, and the adequate predicate. For example:

(SORT-A-LIST *list-of-numbers* #'<)

would sort a list of numbers in increasing order, and

(SORT-A-LIST *list-of-characters* #'CHAR>)

would sort a list of characters in decreasing order.

A first version of this new function could be written like the following:

```
(DEFUN SORT-A-LIST (L PREDICATE)
      (INSERT PREDICATE (CAR L) (SORT-A-LIST (CDR L))))

(DEFUN INSERT (PREDICATE ELE L) (COND
      ((NULL L) (LIST ELE))
      ((APPLY PREDICATE (LIST ELE (CAR L))) (CONS ELE L))
      (T (CONS (CAR L) (INSERT PREDICATE ELE (CDR L))))))
```

Here we introduced an additional argument. This argument, PREDICATE, will be bound to a function and then, inside the INSERT function, it will be applied to the arguments ELE and (CAR L).

Below are some possible calls to this new version of our sorting function:

```
? (SORT-A-LIST '(1 2 3 3 2 1 0 -1) #'>)
= (3 3 2 2 1 1 0 -1)
? (SORT-A-LIST '(#\d #\c #\b #\a) #'CHAR<)
= (#\a #\b #\c #\d)
? (SORT-A-LIST '(1 2 3 3 2 1 0 -1) #'<)
= (-1 0 1 1 2 2 3 3)
```

We had already met the function APPLY in Chapter 12. There is another function, similar to apply, the function FUNCALL. The difference between FUNCALL and APPLY is that while APPLY expects always *two* arguments: a function to compute, and a list of its arguments, FUNCALL takes as arguments a function to compute and the arguments to this function. For example, the following two expressions are computing exactly the same:

```
(APPLY 'CONS '(A (B C D)))
(FUNCALL 'CONS 'A '(B D C))
```

and, if the value of the symbol F is CONS, the following two expressions are still computing the same value:

```
(APPLY F '(A (B C D)))
(FUNCALL F 'A '(B C D))
```

The facility for computing a function can be used to construct functions with functional arguments. Below is a LISP function which applies function F *twice* to its argument X:

```
(DEFUN TWICE (F X) (FUNCALL F (FUNCALL F X)))
```

Here are a few examples of calls:

```
(TWICE'1+ 2)              → 4
(TWICE 'CAR '((A B C)))   → A
(TWICE 'FACTORIAL 3)      → 720
```

This technique can be used to define very useful functions: we often have to use functions which apply a function to the successive elements of a list, as is the case in the following example:

```
(DEFUN PLUS-ONE (L)
       (IF (NULL L) ()
           (CONS (1 + (CAR L)) (PLUS-ONE (CDR L)))))
```

which returns a copy of the original list with each element incremented by 1. Another example of the same type of program is:

```
(DEFUN LIST-EACH (L)
       (IF (NULL L) ()
           (CONS (LIST (CAR L)) (LIST-EACH (CDR L)))))
```

which is a function returning a copy of the list with each element enclosed in a further pair of parentheses.

This type of function appears so often that it is worth constructing a function encapsulating the basic pattern. Here it is:

```
(DEFUN MAPCAR (F L)
       (AND L
            (CONS (FUNCALL F (CAR L))
                  (MAPCAR F (CDR L)))))
```

Using this function MAPCAR, the two functions above can be written more simply as follows:

```
(DEFUN PLUS-ONE (L) (MAPCAR #'1+ L))

(DEFUN LIST-EACH (L) (MAPCAR #'LIST L))
```

MAPCAR is therefore a function allowing a function to be applied to the successive elements of a list, returning as result the list of the successive results.

Take a moment to appreciate the power of this function: by taking advantage of the possibility to pass functions as arguments we constructed a function which totally separates the control structure of the real computation. This function may be used for many different applications, each one specified only by the adequate functional argument. Here are some further possible applications of this function MAPCAR:

```
? (MAPCAR 'FACTORIAL '(1 2 3 4 5))
= (1 2 6 24 120)
? (MAPCAR 'NUMBERP '(1 A B 2 3 C))
= (T NIL NIL T T NIL)
```

```
? (MAPCAR 'SQRT '(10 20 30 40 50))
= (3.162278S0 4.472136S0 5.477226S0 6.324555S0 7.071068S0)
? (MAPCAR 'ABS '(-200 100 -3 -4 5 6))
= (200 100 3 4 5 6)
? (MAPCAR 'RANDOM '(100 200 300 400))
= (59 158 6 162)
```

Actually, LISP has a standard function MAPCAR which is still more powerful: if the function we wish to apply repeatedly requires more than one argument, we can supply more than one argument list to MAPCAR. For example, suppose that we want to construct a list which contains the CONS of 1 and 2, the cons of 3 and 4, the cons of 5 and 6, and the cons of 7 and 8. We can do this, in supplying to MAPCAR the function CONS, the list (1 3 5 7) and the list (2 4 6 8). Here is what this gives:

```
? (MAPCAR #'CONS '(1 3 5 7) '(2 4 6 8))
= ((1 . 2) (3 . 4) (5 . 6) (7 . 8))
```

MAPCAR evaluates *each* argument. Then it applies the first argument to the successive CARs of the other arguments. In this way, CONS gets first to 1 and 2, then to 3 and 4, then to 5 and 6, and finally to 7 and 8. MAPCAR collects all the successive results of these computations and returns them all in a list, when finished.

Below are a few additional examples of the use of MAPCAR:

```
? (MAPCAR #' + '(1 2 3 4 5) '(10 20 30 40 50))
= (11 22 33 44 55)
? (MAPCAR #'APPEND '((A B) (C D) (E F)) '((1 2) (3 4) (5 6)))
= (A B 1 2) (C D 3 4) (E F 5 6))
? (SETQ LIST1 '((A B) (C D) (E F)) LIST2 '(1 2 3))
= (1 2 3)
? (MAPCAR #'RPLACA LIST1 LIST2)
= ((1 B) (2 D) (3 F))
? LIST1
= ((1 B) (2 D) (3 F))
```

Let us now turn to the various types of function calls. We already know two: explicit calls and calls by evaluation of a variable in the functional position. Of course, we are not limited in LISP to having atoms in the functional position: we could place any expression there, with the proviso that their evaluation must give a functional object. Here are some examples:

```
(FUNCALL (CAR '(1+ 1-)) 3)                          → 4
(FUNCALL (LET ((F 'CAR)) F) '(A B C))               → A
(FUNCALL (TWICE 'CAR '((FACTORIAL)(+)(-))) 6)       → 720
((LAMBDA (X) (LIST X X)) 2)                         → (2 2)
```

Look at the last example closely: It uses a λ-expression. Such an expression is a function without a name whose evaluation returns the λ-expression itself.

```
(LAMBDA (X) (LIST X X))
```

is a nameless function with one argument, called X, and whose body is (LIST X X). We can call it simply by putting this λ-expression in the functional position. This is very similar to our LET function which combines the definition of a λ-expression with its call. In fact, LET can be defined in terms of LAMBDA as follows:

```
(DEFMACRO LET L
   `((LAMBDA ,(MAPCAR 'CAR (CAR L)) ,@(CDR L))
     ,@(MAPCAR 'CADR (CAR L))))
```

It is therefore possible to use functions more complex than standard functions in calls to MAPCAR. Below is a function which returns the list of all the sub-lists of a list, after having eliminated all occurrences of the atom A:

```
(DEFUN REM-A (L)
      (MAPCAR '(LAMBDA (X) (DELQ 'A X)) L))
```

If we call:

```
(REM-A '((A B C) (A C B) (B A C) (B C A) (C A B) (C B A)))
```

the result is the list:

```
((B C) (C B) (B C) (B C) (C B) (C B))
```

and if we want to obtain a list without repetitions, we simply need the following two functions:

```
(DEFUN REM-A-NO-REP (L) (REM-A-AUX (REM-A L)))

(DEFUN REM-A-AUX (X)
        (IF (NULL X) ()
            (IF (MEMBER (CAR X) (CDR X))
                (CONS (CAR X)
                  (REM-A-AUX (DELETE (CAR X) (CDR X))))
                (CONS (CAR X) (REM-A-AUX (CDR X))))))
```

This, using another iteration function MAPC, can be simplified to:

```
(DEFUN REM-A-NO-REP (L)
   (LET ((AUX NIL))
      (MAPC '(LAMBDA (X)
                 (IF (MEMBER X AUX) NIL
```

```
                    (SETQ AUX (CONS X AUX))))
                (REM-A-L))
        AUX))
```

Function MAPC is a simplification of MAPCAR defined as follows:

```
(DEFUN MAPC (F L) (COND
     ((NULL L) NIL)
     (T (FUNCALL F (CAR L)) (MAPC F (CDR L)))))
```

The call:

```
(REM-A-NO-REP '((A B C) (A C B) (B A C) (B C A) (C A B) (C B A)))
```

now returns:

```
((C B) (B C))
```

Functions of the MAPxxx type, which apply a function to the successive elements of a list, open the way to a whole series of conceptually simple algorithms (which does not necessarily mean they are efficient). So, for example, to find all the possible permutations of the elements of a list, a simple algorithm is to take one element after the other in a list and to distribute it in all possible positions within the list *without* that element. This gives the program:

```
(DEFUN PERMUT (L) (PERMUT1 L NIL)

(DEFUN PERMUT1 (L RES)
        (LET ((X NIL))
             (MAPC
               '(LAMBDA (ELT)
                  (SETQ X (DELQ ELT L))
                  (IF X (PERMUT1 X (CONS ELT RES))
                     (PRINT (CONS ELT RES)))) L)))
```

The eight queens problem is very similar to that of finding permutations. The object is to put eight queens on a chessboard in such a way that no queen could take another.

Remember that a chessboard contains 64 squares distributed over eight columns and eight rows. A queen is a piece which can take other pieces on the same row, column or diagonal.

Obviously, a solution must have just one queen per row and per column. We can therefore simplify our representation, and instead of using a two-dimensional array for the chessboard, we can represent it as a list of eight elements, where the position of an element represents the column (so the first element represents the first column, the second element the second column etc) and where the element indicates the row within the corresponding column where the queen should be placed. For example, the list:

(1 7 5 8 2 4 6 3)

represents the following position:

R							
				R			
							R
					R		
		R					
						R	
	R						
			R				

With this representation, the problem is simply one of finding all the permutations of the numbers 1 to 8 satisfying certain constraints. The following program solves the eight queens problem:[3]

```
(DEFUN QUEEN (LST) (QUEEN-AUX LST 1 NIL NIL NIL NIL))

(DEFUN QUEEN-AUX (LST COLUMN RESUL DIAG1 DIAG2 AUX3)
      (MAPC
           '(LAMBDA (X)
                 (SETQ AUX3 (DELQ X LST))
                 (IF AND (NULL (MEMQ (-X COLUMN) DIAG1))
                         (NULL (MEMQ (+X COLUMN) DIAG2)))
                    (IF (NULL AUX3) (PRINT (CONS X RESUL))
                         (QUEEN-AUX AUX3
                                (1+ COLUMN)
                                (CONS X RESUL)
                                (CONS (-X COLUMN) DIAG1)
                                (CONS (+X COLUMN) DIAG2))))) L))
```

A call then takes the form

```
(QUEEN '(1 2 3 4 5 6 7 8))
```

Function LIT is similar to MAPCAR. Here it is:

```
(DEFUN LIT (1ST END FUNCTION)
     (IF LST
         (FUNCALL FUNCTION (NEXTL LST) (LIT LST END FUNCTION))
         END))
```

With this function, some of the functions examined in this book can be considerably simplified. The function APPEND, for example, can be rewritten as follows:

---

[3] A very good description of this problem is given in Jacques Arsac's book *La construction de programmes structures*.

```
(DEFUN APPEND (L1 L2) (LIT L1 L2 'CONS))
```

nd the definition of function MAPCAR becomes:

```
DEFUN MAPCAR (F L) (LIT L NIL'(LAMBDA (X Y)(CONS (FUNCALL F X) Y))))
```

unction MUL below calculates the product of the numbers in the list given as rgument:

```
(DEFUN MUL (L) (LIT L 1 '*))
```

17.3 SOME MORE ITERATION FUNCTIONS

s well as the facilities for repetition provided by recursion, Le_LISP has a number f *iteration functions*: WHILE, UNTIL and DO. (Common-LISP has, in the standard ersion, only the iterator DO). Function WHILE is defined as follows:

(WHILE *test body*) → NIL
body is repeatedly executed for as long as the evaluation of *test* gives a value other than NIL

Here is function REVERSE written using WHILE:

```
(DEFUN REVERSE (L)
     (LET ((RES NIL))
       (WHILE L (SETQ RES (CONS (NEXTL L) RES)))
       RES))
```

nd function APPEND could be written:

```
(DEFUN APPEND (L1 L2)
        (LET ((AUX (REVERSE L1)))
              (WHILE AUX (SETQ L2 (CONS (NEXTL AUX) L2)))))
        L2)
```

Function UNTIL is very similar to WHILE; it is defined as follows:

(UNTIL *test body*) → NIL
body is repeatedly executed until the evaluation of *test* gives a value other than NIL.

Using UNTIL, our REVERSE function becomes:

```
(DEFUN REVERSE(L)
        (LET ((RES NIL))
              (UNTIL (NULL L) (SETQ RES (CONS (NEXTL L) RES)))
              RES))
```

Naturally, if function WHILE or UNTIL is missing in your version of LISP, you can define them as macros.

The technique described at the beginning of this chapter, i.e. repetition by sequential evaluation of a circular list, can be used within the definition of macro-functions. This would give the following definition for the WHILE function:

```
(DEFMACRO WHILE (&WHOLE ALL &REST CALL)
   `(IF (NULL ,(CAR CALL)) ()
       (PROGN ,@(CDR CALL) ,ALL)))
```

Note that in this macro we have used a new keyword: &WHOLE. The parameter associated to this keyword, here this is the parameter ALL, is bound to the entire macro call. For example, if we call the macro WHILE like:

```
(WHILE (> N 0) (PRINT (SETQ N (1- N))))
```

the parameter ALL will be bound to this entire expression, that is to this list representing the call of WHILE. This is what permits us, in the last line, to construct a circular list. The expression ,ALL will be replaced by a pointer towards the structure itself.

MACLISP, a LISP dialect developed at MIT, introduced the repetition function DO which is defined as follows:

(DO ((*var*$_1$ *init repeat*) . . . (*var*$_n$ *init repeat*))
    (*end-test output-value*)
    body)

This repetition function has three parts:

1. A variable declaration list. As in a LET statement, this gives the initial value of each variable, but also (and this is the original aspect of the DO function) an expression to compute the new value for each iteration.
2. A clause in COND style, where the CAR is a test and the rest a sequence of instructions to evaluate if the test returns true. This test determines the halting point for iteration.
3. A body, containing the sequence of instructions to evaluate on each iteration.

By way of example, here is function COUNT which prints all numbers from zero to N:

```
(DEFUN COUNT (N)
       (DO ((I 0 (1+ I)))
           ((= I N) (TERPRI) 'FINISHED)
           (PRINT I)))
```

Function REVERSE can be written using DO as follows:

```
(DEFUN REVERSE (L)
          (DO ((RES NIL (CONS (CAR L) RES))
               (L L (CDR L)))
              ((NULL L) RES)))
```

n this form, the 'body' part of the DO loop is empty: everything happens during the teration itself.

Here is a possible definition of the DO macro function

```
(DEFMACRO DO CALL
   (LET ((INIT (CAR CALL))
         (STOPTEST (CAAR (CDR CALL)))
         (RETURN-VALUE (CDAR (CDR CALL)))
         (BODY (CDDR CALL)))
      '((LAMBDA ,(MAPCAR 'CAR INIT)
           (WHILE (NULL ,STOPTEST)
                  ,@BODY
                  ,@(MAPCAR '(LAMBDA (X)
                                 '(SETQ ,(CAR X) ,(CADDR X)))
                             INIT))
           ,@RETURN-VALUE)
         ,@(MAPCAR 'CADR INIT))))
```

The function REVERSE defined using the DO macro-function will therefore expand into:

```
(DEFUN REVERSE (L)
        ((LAMBDA (RES L)
           (WHILE (NULL (NULL L)) (SETQ RES (CONS (CAR L) RES))
                  (SETQ L (CDR L)))
           RES)
         NIL L))
```

Here only the macro DO was expanded. When we also expand the macro WHILE, this will finally expand into:

```
(DEFUN REVERSE (L)
      ((LAMBDA (RES L)
          (IF (NULL (NULL (NULL L))) NIL
             (PROGN
               (SETQ RES (CONS (CAR L) RES))
               (SETQ L (CDR L))
               (WHILE (NULL (NULL L)) (SETQ RES (CONS (CAR L) RES))
                      (SETQ L (CDR L)))))
        RES)
      NIL L))
```

Two remarks about this function:

1. Definitely, this is not the most beautiful function definition we've ever seen: the predicate looks somewhat awkward. Nevertheless, the original form is very readable, and every good compiler will optimize the double call of NULL.
2. The remaining WHILE inside the function is just a way to express the circular structure. In reality, this is not a call to WHILE, but a pointer back to the IF function call.

17.4 EXERCISES

1. Write a function MAPS, which applies a function to each sub-structure in a list. Here is an example of its use:

```
? (MAPS '(A (B C) D) 'PRINT)
(A (B C) D)
A
((B C) D)
(B C)
B
(C)
C
(D)
D
= NIL
```

2. Write the macro-function UNTIL.
3. Write function FACTORIAL using the DO function.

18 PATTERN MATCHING (PART TWO)

In Chapter 14 we built up a pattern matching function called MATCH, allowing us to compare a list with a list *model* or *pattern*.

The pattern may contain three special symbols:

1. The $ sign represents any single element.
2. The & sign represents any series of elements, including none at all.
3. The % sign indicates a choice between several given elements.

So, for example, the pattern ($ $ $) corresponds to any list containing exactly three elements, pattern ((%A B C) $) to any list with exactly two elements the first of which is A, B or C and pattern (START & END) to any list starting with element START and ending with element END.

This pattern matching function is not complete. Sometimes we will be interested not only in the structure of a list, but also in the element which appears in the position occupied by one of the special signs. For example, when we compare pattern ($ ON A) with a set of data, we may perhaps want to know what appears in the position occupied by the $ sign.

Let us enhance the syntax of our patterns to allow pattern operators to be followed by a variable name, so that instead of writing:

($ on A)

we could write:

($X on A)

We shall in addition enhance our pattern matching function so that it not only indicates that a particular data item corresponds to the pattern, but also to indicate what element occurs in the position occupied by the variable.

For example, if we match

($X ON A)

with the data

(D ON A)

the pattern matching function returns T, i.e. it tells us that the data item matches the pattern, but also tells us that the $X position is occupied by the element D.

Clearly, two problems are posed. First of all, we have to change our function MATCH so that it returns *two* values: an indication of whether pattern matching has succeeded or failed, and if so the values linked to the different variables in the pattern. This means we have to find a way of expressing links. There are two possibilities: either we can operate as LISP does, i.e. by changing the C-values of variables (using function SET or SETQ), or by returning a list explicitly giving the names of variables followed by their values, i.e. a list of the form:

(($variable_1$. $value_1$) . . . ($variable_n$.$value_n$))

Such lists are called *A-lists* or *association lists*. The earliest implementations of LISP used such lists as *environments*, i.e. as a way of storing current and previous bindings.

Consequently, the result of pattern matching as shown above should give the result:

(T ((X . D)))

where the T at the beginning of the list indicates that pattern matching has succeeded, and the second element of the list is the A-list expressing the fact that X is bound with D.[1] To start on the process of defining our new patterns, let us look at a few examples of matching:

- (MATCH '(A B C D) '(A B C D)) → (T NIL)
  If the filter contains no special sign, pattern matching will only work for a data item which is EQUAL to the pattern. The result is a list, whose CAR is the pattern matching success indicator T and whose CADR is the empty A-list, since no binding is necessary.
- (MATCH '($X B C $-) '(A B C D)) → (T ((X . A)))
  The pattern beginning with the $ symbol must correspond with a single element in the data item. If the $ is followed by an atom other than minus, this atom is bound to the corresponding element in the data item. The special pattern $- shows that a corresponding element must be present in the data item, but without its being necessary to establish a link. The last pattern acts in the same way as the $ sign in the pattern matching function from Chapter 14.
- (MATCH '($X $Y $Z $X) '(A B C A)) → (T ((X . A) (Y . B) (Z . C)))
  Repetition of the same pattern variable within a pattern indicates equal elements.

---

[1] If a *shallow* link is required, i.e. a link which modifies the C-values of variables, we only need to operate on the A-list part of the result as follows:

(MAPC (LAMBDA (X) (SET (CAR X) (CDR X))) *A-list*)

The pattern above will match a data item containing four elements whose first and fourth elements are identical.

- (MATCH '($X &- $X) '(A B C A)) → (T ((X . A)))
  This pattern describes any list beginning and ending with the same element. The pattern &- corresponds to the & pattern in our match function from Chapter 14. It matches a segment of any length *without* forming a link.
- (MATCH '($X $Y $X) '(A B C A)) → (T ((X . A) (Y B C)))
  The difference between this pattern and the preceding one is that the matched segment is associated with variable Y. This variable is therefore bound to the list (B C). (Remember that the CDR of each pair in the A-list contains the value of the variable in the CAR of the pair. This is why we have the pair (Y B C) in the result A-list).
- (MATCH '(%(A B C) &-) '(A B C A)) → (T NIL)
  As before, the special sign % indicates a multiple choice. This pattern therefore fits lists of any length beginning with the atom A, B or C.

We shall now make the necessary modifications to our pattern matching program. Here is its original form:

```
(DEFUN MATCH (PATTERN DATA)
     (LABELS ((CHOICE (AUX PATTERN DATA) (COND
                    ((NULL AUX) NIL)
                    ((MATCH (CAR AUX) (CAR DATA))
                              (MATCH (CDR PATTERN) (CDR DATA)))
                    (T (CHOICE (CDR AUX) PATTERN DATA)))))
       (COND
               ((ATOM PATTERN)(EQUAL PATTERN DATA))
               ((EQUAL (CAR PATTERN) '|&|) (COND
                    ((NULL (CDR PATTERN)) T)
                    ((MATCH (CDR PATTERN) DATA))
                    (DATA (MATCH PATTERN (CDR DATA)))))
               ((ATOM DATA) NIL)
               ((EQUAL (CAR PATTERN) '|$|)
                    (MATCH (CDR PATTERN) (CDR DATA)))
               ((AND (CONSP (CAR PATTERN))
                (EQUAL (CAAR PATTERN) '|%|))
                    (LET ((AUX (CDR (CAR PATTERN))))
                           (CHOICE AUX PATTERN DATA)))
               ((MATCH (CAR PATTERN) (CAR DATA))
                      (MATCH (CDR PATTERN) (CDR DATA))))))
```

First, we have to produce the A-list. We shall therefore change the beginning of the definition of MATCH to:

```
(DE MATCH (PATTERN DATA)
     (LET ((ALIST NIL)) . . .
```

The function's task can be split into two: the actual work will be carried out by an auxiliary function, while MATCH will simply launch this auxiliary function and print the result.

```
(DEFUN MATCH (PATTERN DATA)
    (LET ((ALIST NIL))
        (DECLARE (SPECIAL ALIST))
        (IF (MATCH-AUX PATTERN DATA) (LIST T ALIST) NIL)))
```

Here we DECLAREd the variable ALIST as a SPECIAL variable. Why? In LISP normally we don't need to declare any variable, except to tell some facts, the compiler can't deduce, to the compiler. For example, we might want to tell the compiler that a given variable will only take integer values. This would permit us to compile more efficiently. But here we are not concerned with compilation; this will be addressed in a second volume.

The only declaration which has an effect on interpretation, standard execution of LISP programs, is the declaration of a variable as *special*. A variable to be special means that this variable will follow non-standard access mechanisms: Normally a variable refers either to the local definition of this variable, that is: it is defined inside the block of text (the function, the lambda expression) where it is referenced, or it is a global variable, that is: a variable defined in the top level loop of LISP. This is called *lexical scoping* of variables.

In Common-LISP, it is, however, possible to have variables with *dynamic scoping*. *Dynamic scoping* means that a variable is defined not only in the text where it is defined but also in *all* functions which are dynamically called from the part inside which the variable is declared. That is, it is possible to have variables that are not strictly local to a function. Since this type of variables is not typical, these variables are, in Common-LISP, called *special* variables, and they have to be declared as such.

This way of scoping, i.e. *dynamic* scoping, is the standard scoping mechanism used in most LISP systems prior to Common-LISP. So, if you are using some other LISP you should first examine the scoping mechanism.

Note finally, that the problem of determining if a variable is dynamically or lexically scoped is only critical in occurrences of variables inside functions (or lambda expressions) where they are not part of the list of parameters. Classically, such occurrences are called *free* occurrences. If this is the case, the variable refers either to the most recently created special variable of this name, or, if no special variable of this name exists, to the global variable of this name.

To summarize: a variable that is not declared special, can only be referred to inside the function where it is defined. A special variable can be referred to in any function that is called (directly or indirectly) by the function where this variable has been defined.

Here is a little example: suppose that you have the following program:

```
(SETQ X 'X Y 'Y Z 'Z)

(DEFUN FOO (X Y Z)
   (DECLARE (SPECIAL Z))
                  (BAR X))

(DEFUN BAR (X)
   (FORMAT T "~%x = ~ A y = ~A z= ~A" X Y Z))
```

This program uses three global variables: X, Y and Z. The global variables are all initialized to their respective names.

The function FOO defines two *lexical* variables: X and Y, and one *special* variable: Z.

The function BAR defines one *lexical* variable: X. It contains *free* occurrences of the variables Y and Z.

Let us call FOO:

```
? (FOO 1 2 3)
x= 1 y= Y z= 3
```

Well, from the result we can conclude that in the function BAR the free occurrence of Y refers to the *global* variable of that name, and the free occurrence of Z refers to its most recent *special* definition.

Let us return to our pattern matcher. In this new version of MATCH, we are declaring the variable ALIST to be *special* because we want *this* variable ALIST to be known by all other functions called by MATCH or MATCH-AUX.

Now we have to rewrite the old MATCH function, now called MATCH-AUX. There is nothing to change in the first two clauses. The modifications concern the handling of the special signs $ and &.

We shall first deal with the difficulty of separating the sign ($ or &) from the variable name that follows it. If we use function EXPLODECH to test the first character of an atom, we have to do it with *all* atoms contained in the pattern. This takes too long. Let us therefore define the characters $, & and % as macro-characters. This means that as soon as the special signs have been read they will be separated from variable names following them. Here are the definitions of three such characters first in Le_Lisp:

```
(DMC |$| () (CONS '|$| (READ)))
(DMC |&| () (CONS '|&| (READ)))
(DMC |%| () (CONS '|%| (READ)))
```

and then in Common-LISP:

```
(SET-MACRO-CHARACTER   #\$
          #'(LAMBDA (STREAM CHAR) (CONS '|$| (READ))))
(SET-MACRO-CHARACTER   #\&
          #'(LAMBDA (STREAM CHAR) (CONS '|&| (READ))))
(SET-MACRO-CHARACTER   #\%
          #'(LAMBDA (STREAM CHAR) (CONS '|%| (READ))))
```

Consequently, the pattern:

```
($X &Y %(A B C))
```

will be translated into:

```
(($ . X) (& . Y) (% A B C))
```

and the tests for the presence of special signs only takes place if the pattern starts with a sub-list. We can therefore introduce a further clause testing to see whether the pattern begins with an atom. If it does, we know that it must be a constant and we need only compare that constant with the first element in the data item and then launch the pattern matching process again for the rest of the two lists. This gives us:

```
((ATOM (CAR PATTERN))
        (AND (EQUAL (CAR PATTERN)(CAR DATA))
             (MATCH-AUX (CDR PATTERN) (CDR DATA))))
```

In all other cases, the pattern begins with a sub-list. This sub-list may correspond with one of the special characters. We shall consider all possible cases and start with the character $:

```
((EQUAL (CAAR PATTERN) '|$|). . .)
```

Three sub-cases may occur:

1. The data is the empty list. In this case, matching fails and we can exit returning the value NIL.
2. The dollar sign is followed by a '−' sign. We are therefore in the same position as with the old MATCH function: we need only compare the rest of the pattern with the rest of the data.
3. The dollar sign is followed by the name of a variable. Two cases have to be distinguished:
   (a) The variable has not yet been bound on the A-list. The variable must be bound to the corresponding element in the data and we then go on to compare the rest of the pattern with the rest of the data.
   (b) The variable has already been bound on the A-list. We compare the data with the pattern constructed by replacing the first element by the value of the variable.

Let us first build a small auxiliary function ASSOC which searches through the A-list to see if a given variable has already been bound or not. If it has been bound, ASSOC returns the pair (*variable . value*):

```
(DEFUN ASSOC (VARIABLE ALIST)(COND
          ((NULL ALIST) NIL)
          ((EQUAL (CAAR ALIST) VARIABLE) (CAR ALIST))
          (T (ASSOC VARIABLE (CDR ALIST)))))
```

This function appears as a standard function in most LISP systems.

Now we can write the clause for the $ sign:

```
((EQUAL (CAAR PATTERN) '|$|) (COND
      ((NULL DATA) NIL)
      ((EQUAL (CDAR PATTERN) '|-|)
            (MATCH-AUX (CDR PATTERN) (CDR DATA)))
      ; check if there is already a binding
      ((ASSOC (CDAR PATTERN) ALIST)
            (MATCH-AUX
                  (CONS (CDR (ASSOC (CDAR PATTERN) ALIST))
                        (CDR PATTERN))
                  DATA))
      ; if there is no binding
      (T (SETQ ALIST
            (CONS (CONS (CDAR PATTERN) (CAR DATA)) ALIST))
        (MATCH-AUX (CDR PATTERN) (CDR DATA)))))
```

Clearly, this clause deals in turn with each of the cases we have been discussing. There is just one defect still: the binding of a variable is computed twice. To get round this problem, we can include another variable in our function MATCH-AUX, which we shall call AUX for "auxiliary", and use this variable as a temporary store for a binding. This means we have to modify function MATCH-AUX as follows:

```
(DEFUN MATCH-AUX (PATTERN DATA)
   (LET ((AUX NIL))
        (DECLARE (SPECIAL AUX))
```

and rewrite the second to last clause:

```
((SETQ AUX (ASSOC (CDAR PATTERN) ALIST))
   (MATCH-AUX (CONS (CDR AUX) (CDR PATTERN)) DATA))
```

Now we shall consider the case of the pattern %, i.e. the indicator for an explicit multiple choice. What is different from the old function is the need to store the state of the A-list during successive trails; this allows us to introduce patterns within the choice list. For example, the pattern:

```
(% (DO RE (MI $-) (FA $- $-)) &-)
```

describes any list starting either with atom DO, atom RE, a list containing two elements beginning with MI, or a list of three elements beginning with the atom FA.

Here is the clause corresponding to the % sign:

```
((EQUAL (CAAR PATTERN) '|%|)
        (CHOICE (CDAR PATTERN) PATTERN DATA ALIST))
```

and the little function doing the effective choice:

```
(LABELS ((CHOICE (AUX PATTERN DATA ALIST2) (COND
        ((NULL AUX) (SETQ ALIST ALIST2) NIL)
        ((MATCH-AUX (CONS (CAR AUX) (CDR PATTERN)) DATA) T)
        (T (CHOICE (CDR AUX) PATTERN DATA ALIST2)))))
```

Next comes the difficult case of segment variables, i.e. variables preceded by the special sign '&'. Three cases have to be distinguished.

1. The variable has already been bound. The variable must therefore be replaced, in the pattern, by its value, and the pattern obtained in this way must then be compared with the data.
2. The rest of the pattern is empty. We then have to bind the variable (unless it is the special sign −) with the data; the matching operation ends in success.
3. Otherwise, the variable must be bound (again, unless it is the special sign −) with the empty segment NIL, and a check must be made to see whether the rest of the pattern can be superimposed on the data. If this is not the case, the first element of the data must be added to the value of the variable, and the rest of the pattern has to be compared with the rest of the data and so on until a definite success or failure is reached.

This difficult clause can be written with the auxiliary function SEQUENCES and the following clause:

```
(SEQUENCES (ALIST2 AUX PATTERN DATA) (COND
  (DATA (IF
          ; carry out the successive comparisons
          (MATCH-AUX (CDR PATTERN) DATA) T
          (PROGN (SETQ ALIST ALIST2)
               (IF AUX (NEXTL DATA)
                    ; add to variable's binding
                    (NCONC (CAR ALIST) (LIST (NEXTL DATA))))
               (SEQUENCES ALIST AUX PATTERN DATA))))
  ; total failure: restore the alist
  (T (IF AUX (NEXTL ALIST)) NIL))))
```

```
((EQUAL (CAAR PATTERN) '|&|) (COND
        ; is a value already present?
        ((SETQ AUX (ASSOC (CDAR PATTERN) ALIST))
             ; yes, compare it with the data value
             (MATCH-AUX (APPEND (CDR AUX) (CDR PATTERN)) DATA))
        ; are we at the end of the pattern?
        ((NULL (CDR PATTERN))
                  ; success
                  (IF (EQUAL (CDAR PATTERN) '|-|) NIL
                      (SETQ ALIST
                           (CONS (CONS (CDAR PATTERN) DATA) ALIST))) T)
        (T (IF (SETQ AUX (EQUAL (CDAR PATTERN) '|-|)) NIL
               (SETQ ALIST (CONS (LIST (CDAR PATTERN)) ALIST)))
           (SEQUENCES ALIST AUX PATTERN DATA))))
```

Only the final case remains: the pattern begins with an ordinary sub-list. As in the old version, we have to compare the first element of the pattern with the first element of the data, and, if that works, compare the rest of the pattern with the rest of the data.

Here is a series of tests with the new MATCH function:

```
(MATCH '(A B C)'(A B C))                          → (T NIL)
(MATCH '($- B C)'(A B C))                         → (T NIL)
(MATCH '($- B C $-)'(A B C))                      → NIL
(MATCH '($X B C)'(A B C))                         → (T ((X . A)))
(MATCH '($X B $Y)'(A B C))                        → (T ((Y . C)(X . A)))
(MATCH '($X B $X)'(A B A))                        → (T ((X . A)))
(MATCH '(&- B C)'(A B C))                         → (T NIL)
(MATCH '(&X %(1 2 3) &Y)'(A B C 1 A B C D))       → (T ((Y A B C D) (X A B
                                                     C))
(MATCH '(%(A(A $X)(A $X $-))$X)'((A B C) B))      → (T ((X . B)))
(MATCH '(&X 1 &X)'(A B C 1 A B C D))              → NIL
(MATCH '(&X &X &X)'(A B C A B C A B C))           → (T ((X A B C)))
(MATCH '($A &-($- &B (&C)&D))'(1 2 3 (4 5 (6 7)))) → (T((D)(C 6 7)(B . 5)(A .
                                                     1)))
```

As an example of the use of this pattern matching program, let us take a small program using this algorithm to test whether a given word is a palindrome or not. Remember that a palindrome is a word which reads the same backwards or forwards. For example, the word 'otto' is a palindrome, as are the words 'aha' or 'rotor'.[2]

Here is the start of the program:

```
(DEFMACRO PALINDROME (MOT) '(PAL (EXPLODECH ,MOT)))
```

[2] Georges Perec once wrote a short story several pages long in the form of a palindrome, cf. *Oulipo, la litterature potentielle*, pp. 101–106, idees, Gallimard, 1973.

Clearly, this program simply explodes the word into a sequence of characters allowing the auxiliary function PAL to compare the first and last character, and if they are identical, to compare the penultimate character with the second, and so on, until two different characters are encountered, or a single character, or nothing at all.

To compare the first and last characters, the pattern:

```
($X &MIDDLE $X)
```

seems perfect; moreover, this pattern provides us with the sequence of characters remaining if the characters at the beginning and end are removed:

```
(DEFUN PAL (LIST-OF-CHARACTERS)
   (IF (NULL (CDR LIST-OF-CHARACTERS)) 'YES
      (LET ((AUX (MATCH '($X &MIDDLE $X) LIST-OF-CHARACTERS)))
           (IF (CAR AUX) ;what do we do now?
               'NO))))
```

The problem is what to do if pattern matching works. If this happens, variable AUX contains a list of two elements: the first is the pattern matching success indicator and the second is the A-list resulting from pattern matching. It is this A-list which tells us, through the ordered pair (MIDDLE . *value*), what characters are still to be tested. We then only have to call PAL recursively using the value of middle.

This can be done with the call:

```
(PAL (CDR (ASSQ 'MIDDLE (CADR AUX))))
```

or, more elegantly, by putting the recursive call (PAL MIDDLE) in an environment in which the A-list bindings are temporarily valid.

The following function constructs the appropriate environment:

```
(DEFMACRO LETALL CALL
   (LET ((ALIST (EVAL (CAR CALL))))
     `((LAMBDA ,(MAPCAR 'CAR ALIST) ,@(CDR CALL))
        ,@(MAPCAR '(LAMBDA (X) (LIST 'QUOTE (CDR X))) ALIST))))
```

This macro takes an A-list and a sequence of commands to create an expression in which the sequence of commands is evaluated with the bindings determined by the A-list.

For example, if variable L contains the list:

```
((X A B C) (Y . 1))
```

the call:

```
(LETALL L (PRINT X Y)(CONS Y X))
```

will first print (A B C) 1, and then returns the list (1 A B C). This allows us to complete our function PAL by inserting the following command after the line (IF (CAR AUX):

```
(LETALL (CADR AUX) (PAL MIDDLE))
```

which gives finally:

```
(DEFUN PAL (LIST-OF-CHARACTERS)
  (IF (NULL (CDR LIST-OF-CHARACTERS)) 'YES
    (LET ((AUX (MATCH '($X &MIDDLE $X) LIST-OF-CHARACTERS)))
        (DECLARE (SPECIAL AUX))
        (IF (CAR AUX) (LETALL (CADR AUX) (PAL MIDDLE))
          'NO))))
```

Pattern matching, which lies at the heart of most programming languages developed from LISP, opens the way to programming by means of rules. Such systems interpret a series of ordered pairs of the form ⟨situation actions⟩, by comparing them with a given situation. The interpreter evaluates the action part of the first ordered pair for which the siutaiton described is similar to that in the data.

To illustrate this technique rapidly, we shall reconstruct a short version of the program Eliza, one of the first artificial intelligence programs. This program simulates the behaviour of a psychiatrist. The program is made up of a set of ordered pairs ⟨pattern, action⟩. After each statement by the user (the patient), it compares the sequence of patterns within the statement, and if one of the patterns corresponds, it prints the action part, along with references to parts of the patient's statement. Here is a series of such pairs:

```
pattern              action
(goodbye)            ()
(stop)               ()
(no)                 '(you're very negative)
(&- lisp &-)         '(so you know how to program)
(&- old &-)          '(go further back into your past)
(&- years &-)        '(that's a long time)
($-)                 '(that's a bit curt)
(&- I'm not &l)      '(what makes you think you're not ,@l)
(&- am &l)           '(why do you think you are ,@l)
(&- I don't &-)      '(are you sure)
(&- no &-)           '(you're very negative)
(&- you &-)          '(concentrate on your own problems, please)
(&- me &-)           '(tell me about your family)
(&- I &-)            '(how does that appear)
(&- I have &-)       '(for how long)
```

```
(&- how much &-)          '(how much doesn't matter)
($- $- $-)                '(hmmm)
(&-)                      '(I'm not much help to you)
```

Thanks to this little set of rules, we can already produce a dialogue such as the one below (the program's responses appear in italic type, and the user's in normal type):

(I have been living with Mary for three years now)
(that's a long time)
(not at all)
(hmmm)
(I don't think she loves me)
(are you sure)
(yes)
(that's a bit curt)
(I'm not curt)
(what makes you think you're not curt)
(you shouldn't poke fun at me)
(concentrate on your own problems, please)
etc. . .

Though this opening dialogue is already quite interesting, to make it really stimulating we need to increase the number of ⟨pattern, action⟩ pairs and to improve the program that uses them. Here is an example program (we leave it to you to implement it, improve it and test it):

```
(DEFUN ELIZA ()
   (FORMAT T "~%Hello, what's your problem? ")
   (DO ((SENTENCE (REPLY (ANALYZE (READ))) (REPLY (ANALYZE (READ)))))
       ((NULL SENTENCE) "Goodbye")))

(DEFUN ANALYZE (PROBLEM)
   (LABELS ((INTERN-ANALYZE (PAT-REP MATCH-RESULT)
         (IF (CAR (SETQ MATCH-RESULT (MATCH (CAAR PAT-REP) PROB-
         LEM)))
            (THINKS (CADR MATCH-RESULT) (CDAR PAT-REP))
            (INTERN-ANALYZE (CDR PAT-REP) NIL))))
   (INTERN-ANALYZE LIST-PATTERN-REPLY NIL)))

(DEFUN THINKS (ENV MODEL) (EVAL '(LETALL ',ENV ,MODEL)))

(DEFUN REPLY (ANSWER) (IF ANSWER (FORMAT T "~%Psy: ~A " ANSWER))
                                 ANSWER)
```

The only thing missing is to put the list of patterns and replies into the global variable LIST-PATTERN-REPLY, say as follows:

```
(SETQ LIST-PATTERN-REPLY
      '( ((GOODBYE) . ())
         ((STOP) . ())
         ((NO) . (YOU'RE VERY NEGATIVE))
         ((&- LISP &-) . '(SO YOU KNOW HOW TO PROGRAM)

         ;and so on with all the pairs given above;
```

But let us now return to pattern matching. If we could associate restrictions with our pattern variables, we could considerably simplify the writing of programs such as PAL. If we look again at the concept of a palindrome, another way of defining it is to say that it is a sequence of letters followed by the same sequence of letters reversed, possibly separated by a single letter. This could be represented by the pattern:

(&x &x–*backwards*) or (&X $Y &X–*backwards*)

or, in a more specifically LISP way:

```
(OR (&X (Y (EQUAL X (REVERSE Y))))(&X $M & (Y (EQUAL X (REVERSE Y)))))
```

or, more compactly:

```
(&X &(Z (LE (LENGTH Z) 1)) &(Y (EQUAL X (REVERSE Y))))
```

Introducing this additional modification is not difficult: we simply have to add a module translating the constraint variables into their values, and then evaluate the constraints using the macro-function LETALL. We therefore need to introduce the constraint validity test in the clause handling 'element' variables and 'segment' variables

```
((EQUAL (CAAR PATTERN) '|$|) (COND
    ((NULL DATA) NIL)
    ((EQUAL (CDAR PATTERN) '|-|)
            (MATCH-AUX (CDR PATTERN) (CDR DATA)))
    ; check if the variable has a binding
    ((SETQ AUX (ASSOC (CDAR PATTERN) ALIST))
            (MATCH-AUX
                    (CONS (CDR AUX) (CDR PATTERN)) DATA))
    ; if there is no binding
    (T (SETQ ALIST
            (CONS (CONS (VAR (CDAR PATTERN)) (CAR DATA))
                    ALIST))
      (IF (OR (ATOM (CDAR PATTERN)) (NULL (CDDAR PATTERN))
                    (EVAL '(LETALL ',ALIST ,@(CDDAR PATTERN))))
            (MATCH-AUX (CDR PATTERN) (CDR DATA))
            (PROGN (NEXTL ALIST) NIL)))))
```

Function VAR is needed to find the variable part of the pattern, given that it can now be in the CDR position (if there are no constraints) or in the CADR (if there are constraints):

```
(DEFUN VAR (PATTERN)
  (IF (ATOM PATTERN) PATTERN
     (IF (ATOM (CDR PATTERN)) (CDR PATTERN) (CAR PATTERN))))
```

And the new clause for the segment variable becomes:

```
((EQUAL (CAAR PATTERN) '|&|) (COND
      ; is a value already present?
      ((SETQ AUX (ASSOC (CDAR PATTERN) ALIST))
          ; yes, compare it with the data value
          (MATCH-AUX (APPEND (CDR AUX) (CDR PATTERN)) DATA))
      ; are we at the end of the pattern?
      ((NULL (CDR PATTERN))
               ; success
               (IF (EQUAL (CDAR PATTERN) '|-|) NIL
                  (PROGN (SETQ ALIST
                                 (CONS
                                        (CONS (VAR (CDAR PATTERN))
                                                 DATA)
                                        ALIST))
                             (IF (OR (ATOM (CDAR PATTERN))
                                     (NULL (CDDAR PATTERN))
                                              (EVAL
                                       `(LETALL ',ALIST
                                                 ,@(CDDAR PATTERN))))
                              T (PROGN (NEXTL ALIST) NIL)))))
      (T (IF (SETQ AUX (EQUAL (CDAR PATTERN) '|-|)) NIL
            (SETQ ALIST (CONS (LIST (VAR (CDAR PATTERN))) ALIST)))
         (SEQUENCES ALIST AUX PATTERN DATA))))

(SEQUENCES (ALIST2 AUX PATTERN DATA) (COND
   (DATA
          ; test the constraints
          (IF (OR (ATOM (CDAR PATTERN))
                  (NULL (CDDAR PATTERN))
                  (EVAL `(LETALL ',ALIST ,@(CDDAR PATTERN))))
             ; carry out the successive comparisons
             (SETQ AUX1 (MATCH-AUX (CDR PATTERN) DATA)))
          (IF AUX1 T
             (PROGN (SETQ ALIST ALIST2)
                  (IF AUX (NEXTL DATA)
                       ; add to variable's binding
```

```
                    (NCONC (CAR ALIST) (LIST (NEXTL DATA))))
                 (SEQUENCES ALIST AUX PATTERN DATA))))
      ; total failure: restore the alist
      (T (IF AUX (NEXTL ALIST)) NIL))))
```

With this little modification we can now compare structures with constraints on the pattern variables. Some new patterns of this kind might be:

pattern	*data*	*result*
($X A B $(Y (EQ Y (+ 5 X))))	(5 A B 10)	(T((Y . 10)(X . 5)))
($X A B $(Y (EQ Y) (+ 5 X))))	(5 A B 5)	NIL
($X A B &(Y (EQUAL Y (REVERSE Y))))	(1 2 A B 2 1)	(T ((Y 2 1)(X 1 2))
(&X &(Y (EQUAL Y (REVERSE Y))))	(A B C C B A)	(T((Y C B A)(X A B C))

Only one problem remains in our pattern matching function. Look at the following example:

```
(MATCH '((&X+ &Y) *(&X – &Y)) '((A + B + C) * (A + B – C)))
```

We would expect the result of pattern matching to be:

```
(T ((X A + B)(Y C)))
```

In fact, however, our pattern matching function will return a failure. Why? Take a closer look: the problem with our function is that we are using recursion and *backtracking*. The backtracking instruction is explicit, in the clause concerning segment variables. That clause says we can backtrack (and therefore undo the bindings to the A-list) until we find something that works, or until we have exhausted all the possibilities.

Unfortunately, we forbid backtracking in the last clause: once we have completed comparison of the first element of the pattern, here ($X + $Y), with the first element in the data, here (A + B + C), we can no longer undo the resulting binding, even if we later find, when we compare the respective CDRs, that it is incorrect.

With no more sophisticated control structures,[3] the only solution available to us is to translate patterns and data on entry to function MATCH into linear lists and then to retranslate results on exit.

For example, the pattern:

```
((&X + &Y) * (&X – &Y))
```

can be converted into a linear list as follows:

---

[3] We shall examine such control structures in a further book on LISP.

(⟨⟨ $X + $Y ⟩⟩ * ⟨⟨ $X − $Y ⟩⟩)

and the data item ((A + B + C) * (A + B − C)) can be converted into linear form as follows:

(⟨⟨ A + B + C ⟩⟩ * ⟨⟨ A + B − C ⟩⟩)

In this way, the final clause in our pattern matching function, and therefore the clause which prevents backtracking, can be eliminated since it will no longer be necessary: all patterns and all data will be linear.

The complete pattern matching program, including all the modification, now has the following form:

```
(DEFUN MATCH (PATTERN DATA)
   (LET ((ALIST NIL))
       (DECLARE (SPECIAL ALIST))
       (IF (MATCH-AUX (LINEAR PATTERN) (LINEAR DATA)) (LIST T ALIST)
               NIL)))

(DEFUN MATCH-AUX (PATTERN DATA)
   (DECLARE (SPECIAL DATA)) ; data will be used by DELINEAR
   (LET ((AUX NIL)(AUX1 NIL))
       (DECLARE (SPECIAL AUX AUX1))
   (LABELS (; the function handling multiple choices
          (CHOICE (AUX PATTERN DATA ALIST2) (COND
             ((NULL AUX) (SETQ ALIST ALIST2) NIL)
             ((MATCH-AUX (CONS (CAR AUX) (CDR PATTERN)) DATA) T)
             (T (CHOICE (CDR AUX) PATTERN DATA ALIST2))))
          ; the function handling the sequences
          (SEQUENCES (ALIST2 AUX PATTERN DATA) (COND
             (DATA
                 ; test the constraints
                 (IF (AND (CONSP (CDAR PATTERN))
                          (CDDAR PATTERN))
                    (IF (EVAL `(LETALL ',ALIST ,@(CDDAR PATTERN)))
                          ; carry out the successive comparisons
                          (SETQ  AUX1  (MATCH-AUX  (CDR  PATTERN)
                                                     DATA))  NIL)
                      (SETQ AUX1 (MATCH-AUX (CDR PATTERN) DATA)))
                 (IF AUX1 T
                    (PROGN (SETQ ALIST ALIST2)
                    (IF AUX (NEXTL-DATA)
                          ; add to variable's binding
                          (NCONC (CAR ALIST) (LIST (NEXTL DATA))))
                    (SEQUENCES ALIST AUX PATTERN DATA))))
             ; total failure: restore the alist
             (T (IF AUX (NEXTL ALIST)) NIL))))
```

```
(COND
    ; two end-of-recursion tests:
    ((NULL PATTERN) (NULL DATA))
    ((ATOM PATTERN) (EQUAL PATTERN DATA))
    ; a constant:
    ((ATOM (CAR PATTERN))
        (AND (EQUAL (CAR PATTERN) (CAR DATA)) ; if equality
            (MATCH-AUX (CDR PATTERN) (CDR DATA)))) ; continue
    ; a sequence variable:
    ((EQUAL (CAAR PATTERN) '|&|) (COND
        ; is a value already present?
        ((SETQ AUX (ASSOC (CDAR PATTERN) ALIST))
            ; yes, compare it with the data value
            (MATCH-AUX (APPEND (LINEAR (CDR AUX))
                               (CDR PATTERN)) DATA))
        ; are we at the end of the pattern?
        ((NULL (CDR PATTERN))
                ; then success
                (IF (EQUAL (CDAR PATTERN) '|-|)
                    NIL ; nothing to do if anonymous
                    ; otherwise, create a link
                    (PROGN (SETQ ALIST
                                (CONS
                                    (CONS (VAR (CDAR PATTERN))
                                        (DELINEAR-DATA))
                                    ALIST))
                            ; are there any constraints?
                            (IF (OR (ATOM (CDAR PATTERN))
                                    (NULL (CDDAR PATTERN))
                                    (EVAL ; evaluate them
                                      `(LETALL ',ALIST
                                    ,@(CDDAR PATTERN))))
                                T ; success
                                ; failure: constraint not satisfied
                                (PROGN (NEXTL ALIST) NIL)))))
        (T (IF (SETQ AUX (EQUAL (CDAR PATTERN) '|-|)) NIL
             (SETQ ALIST (CONS (LIST (VAR (CDAR PATTERN))) ALIST)))
           (SEQUENCES ALIST AUX PATTERN DATA))))
        ; atomic data: failure
    ((AND DATA (ATOM DATA)) ())
    ; an element variable
    ((EQUAL (CAAR PATTERN) '|$|) (COND
        ((NULL DATA) NIL) ; no data: failure
        ((EQUAL (CDAR PATTERN '|-| : anonymous?
            (NEXTL DATA) ; advance
            (MATCH-AUX (CDR PATTERN) DATA))
        ; check if the variable has a binding
        ((SETQ AUX (ASSOC (VAR (CDAR PATTERN)) ALIST))
            (MATCH-AUX
```

```
                (IF (ATOM (CDR AUX))
                    (CONS (CDR AUX)(CDR PATTERN))
                    (APPEND (LINEAR (LIST (CDR AUX)))
                                (CDR PATTERN)))
                DATA))
        ; if there is no binding
        (T (SETQ ALIST
                (CONS (CONS (VAR (CDAR PATTERN)) (NEXTL-DATA))
                      ALIST))
           ; are there any constraints
           (IF (OR (ATOM (CDAR PATTERN)) (NULL (CDDAR PATTERN))
                   ; then evaluate them
                   (EVAL `(LETALL ',ALIST ,@(CDDAR PATTERN))))
                ; and continue
                (MATCH-AUX (CDR PATTERN) DATA)
                ; constraints not satisfied: failure
                (PROGN (NEXTL ALIST) NIL)))))
        ; a multiple choice pattern
    ((EQUAL (CAAR PATTERN) '|%|)
                (CHOICE (CDAR PATTERN) PATTERN DATA ALIST)))))))

(DEFUN VAR (PATTERN) ; to find the pattern variable
   (IF (ATOM PATTERN) PATTERN
      (IF (ATOM (CDR PATTERN)) (CDR PATTERN) (CAR PATTERN))))

(DEFUN LINEAR (DATA) (COND ; to convert to linear form
         ((ATOM DATA) DATA)
         ((ATOM (CAR DATA))
            (CONS (CAR DATA) (LINEAR (CDR DATA))))
         ((MEMBER (CAAR DATA) '(|$| |&| |%|))
            (CONS (CAR DATA) (LINEAR (CDR DATA))))
         (T (APPEND (CONS '|<<| (LINEAR (CAR DATA)))
                    (CONS '|>>| (LINEAR (CDR DATA)))))))

(DEFUN DELINEAR (DATA) ; to reconvert from linear form
     (DECLARE (SPECIAL DATA))
     (IF (ATOM DATA) DATA (DELINEAR-DATA)))

(DEFUN DELINEAR-DATA () (COND ; auxiliary for DELINEAR
         ((ATOM DATA) DATA)
         ((EQUAL (CAR DATA) '|<<|)
                (NEXTL DATA)
                (CONS (DELINEAR-DATA) (DELINEAR-DATA)))
         ((EQUAL (CAR DATA) '|>>|)
                (NEXTL DATA) NIL)
         (T (CONS (NEXTL DATA) (DELINEAR-DATA)))))
```

```
(DEFUN NEXTL-DATA () (COND ; to move forward through the linear data
   ((EQUAL (CAR DATA) '|<<|) (NEXTL DATA) (DELINEAR-DATA))
   (T (NEXTL DATA))))

(DEFMACRO NEXTL (VAR) ; to advance one element
     (LET ((VAL (GENSYM)))
          `(LET ((,VAL ,VAR))
               (SETQ ,VAR (CDR ,VAL)) (CAR ,VAL))))

; the macro characters
(SET-MACRO-CHARACTER  #\$
       #'(LAMBDA (STREAM CHAR) (CONS '|$| (READ))))
(SET-MACRO-CHARACTER  #\&
       #'(LAMBDA (STREAM CHAR) (CONS '|&| (READ))))
(SET-MACRO-CHARACTER  #\%
       #'(LAMBDA (STREAM CHAR) (CONS '|%| (READ))))

; and finally, here is the function LETALL again
(DEFMACRO LETALL CALL ; the binding macro
   (LET ((ALIST (EVAL (CAR CALL))))
     `((LAMBDA ,(MAPCAR 'CAR ALIST) ,@(CDR CALL))
        ,@(MAPCAR '(LAMBDA (X) (LIST 'QUOTE (CDR X))) ALIST))))
```

By way of conclusion to the whole book, we shall construct a small rule interpreter using this pattern matching algorithm. This program uses a set of rules of the form:

(*<pattern><action>*)

where the pattern determines *whether* an activity has to be launched, and the *action* part determines *what to do* with the data corresponding to the *pattern*.

If we apply our interpreter to the task of simplifying algebraic expressions, an example pattern might be:

```
($X + 0)
```

and the corresponding action part would be X. This is clearly a rule saying that 0 is a neutral element to the right for addition.

The interpreter has to apply the set of rules until the result of that application is identical to the original expression, i.e. until no further rule can be applied.

Here is the top-level loop of our interpreter:

```
(DEFUN INTERPRETER (EXPR &OPTIONAL EXP1 X) (COND
       ((EQUAL EXP1 EXPR) EXPR)
```

```
        (T (SETQ X (SIMPL EXPR RULES))
           (INTERPRETER X EXPR))))
```

The first clause in this function is the halt clause: the interpreter stops applying the set of rewrite rules when the expression computed by the auxiliary function SIMPL is equal to the expression submitted as data to this function.

The global variable RULES contains the set of rules. If we take the example of a straightforward algebraic simplifier (i.e. one that only works on completely parenthesized algebraic expressions), the set of rules could be the following list:

```
(SETQ RULES '(
        (($(X (NUMBERP X)) + $(Y (NUMBERP Y))) (+ X Y))
        (($(X (NUMBERP X)) * $(Y (NUMBERP Y))) (* X Y))
        (($(X (NUMBERP X)) - $(Y (NUMBERP Y))) (- X Y))
        (($(X (NUMBERP X)) / $(Y (NUMBERP Y))) (/ X Y))
        (($X - 0) X)
        (($X - 0) X)
        ((0 + $X) X)
        ((0 - $X) (LIST '- X))
        (($X * 1) X)
        ((0 * $X) 0)
        (($X * 0) 0)
        ((1 * $X) X)
        (($X / 1) X)
        (($X + $X) (LIST 2 '* X))
        (($X / $X) 1)
        ))
```

But we still need function SIMPL which scans this list of rules and, if one applies, returns the corresponding rewritten form and then re-applies the rules to that form:

```
(DEFUN SIMPL (EXPR RULE &OPTIONAL ALIST) (COND
        ((NULL EXPR) NIL)
        ((ATOM EXPR) EXPR)
        ((CAR (SETQ ALIST (MATCH (CAAR RULE) EXPR)))
                (SIMPL
                    (EVAL `(LETALL ',(CADR ALIST) ,(CADAR RULE)))
                        RULES))
        ((AND RULE (SIMPL EXPR (CDR RULE))))
        (T (CONS (SIMPL (CAR EXPR) RULES)
                (SIMPL (CDR EXPR) RULES)))))
```

And that's all it takes!

Here is a brief interaction with this mini-simplifier:

```
? (INTERPRETER '(4 * (5 + 2)))
= 28
? (INTERPRETER '(4 * (A * 1)))
= (4 * A)
? (INTERPRETER '((A + 0) *((B / B) * 1)))
=A
? (INTERPRETER '((A * (3 * 3)) - ((B + B) / (B - 0))))
= ((A * 9) - (2 * B) / B))
```

If we add the following two rules to our set:

```
(((($X * $Y) / $Y) X)
((($Y * $X) / $Y) X)
```

the last interaction would give us:

```
?(INTERPRETER '((A * (3 * 3)) - ((B + B) / (B - 0))))
=((A * 9) - 2)
```

This interpreter can therefore be quite simply extended by adding new rules. Of course, our rule interpretation program is not very efficient, but it is very general and can be applied to any algorithm expressed by a set of rewrite rules: we only need to change the rule set. Possible improvements would be:

1. A facility to group rules together so that not all rules have to be scanned at each iteration;
2. A facility to modify the order of rules according to the data submitted, to accelerate searching for the applicable rule;
3. A facility allowing several patterns as well as several actions to be included in each rule.

This introduction to LISP programming ends here. We have only seen a very small part of what is possible in programming in general, and in the LISP language in particular. Only the absolutely necessary basics have been described. Now the issue is to practice using the language: write your own programs, read and modify programs written by others.

This book will have fulfilled its aim if it has made you want to program. As with any new technique, it is not by reading a book, however good, that you will become an expert programmer, but by actually writing your own programs, finding the errors you make and putting them right, and by studying just how you can interact with the machine.

19 ANSWERS TO EXERCISES

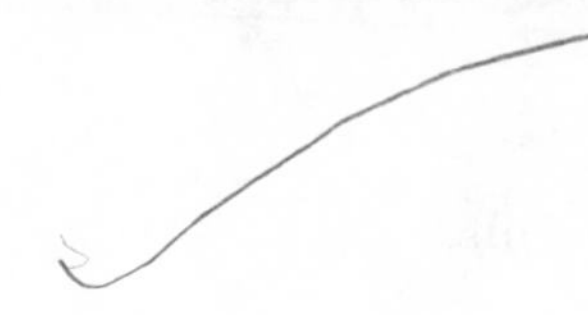

19.1 CHAPTER 1

1. Type determination:

123	numeric atom
(EIN (SCHONES) BUCH)	list
(AHA (AMAZING . . .!))	illegal because of the dots
(((((1) 2) 3) 4) 5)	list
−3Aiii	alpha numeric atom
T	alpha numeric atom
(((((ARRRGH)))))	list

2. Determination of number of elements:

(EIN (SCHONES) BUCH)	three elements: EIN,(SCHONES), BUCH
(((((1) 2) 3) 4) 5)	two elements: ((((1)2)3)4),5
(((((ARRRGH)))))	one element: ((((ARRRGH))))

(UNIX (IS A) TRADEMARK (OF) BELL LABS)
six elements: UNIX, (IS A), TRADEMARK, (OF), BELL, LABS

3. Determination of the maximum depth number:

$(_1EIN(_2SCHONES)_2BUCH)_1$	SCHONES at depth 2
$(_1(_2(_3(_4(_51)_5\ 2)_4\ 3)_3\ 4)_2\ 5)_1$	1 at depth 5
$(_1(_2(_3(_4(_5ARRRGH)_5)_4)_3)_2)_1$	ARRRGH at depth 5
$(_1UNIX\ (_2\ IS\ A)_2\ TRADEMARK\ (_2OF)_2BELL\ \ LABS)_1$	IS, A and OF at depth 2

19.2 CHAPTER 2

1. Combinations of CAR, CDR:

a. (A (B C))
b. (D (E F))
c. (B C)
d. (E F)

e. (NOBODY IS PERFECT)
f. ((CAR A) (CONS A B))

19.3 CHAPTER 3

1. Combinations of CAR and CDR, and CONS:

```
(CAR (CDR (CDR (CDR '(A B C D)))))         → D
(CAR (CDR (CADR (CAR '((A (B C)) E)))))  → C
(CAR (CAR (CAR '(((GOD) STILL) ONE))))→ GOD
(CADR (CAR '(((GOD) STILL) ONE))))       → STILL
```

2. Translation into tree structured form:

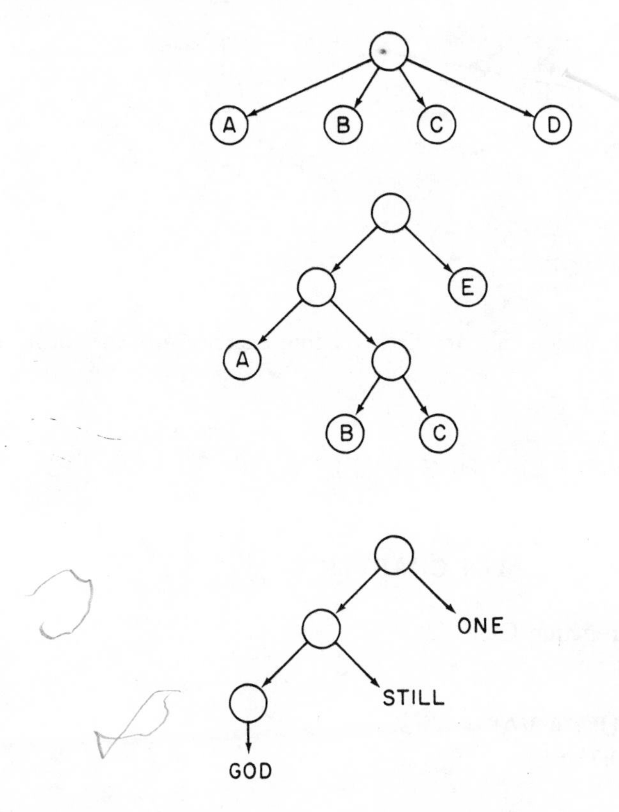

3. SO, ((C D)), ((HELLO) HOW ARE YOU), (JE JE JE BALBUTIE), (C EST SIMPLE)

4. Calculation of tree-operations:

(A)

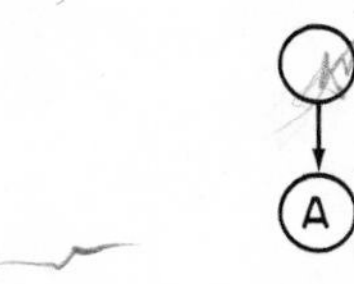

((B C) (D F E))

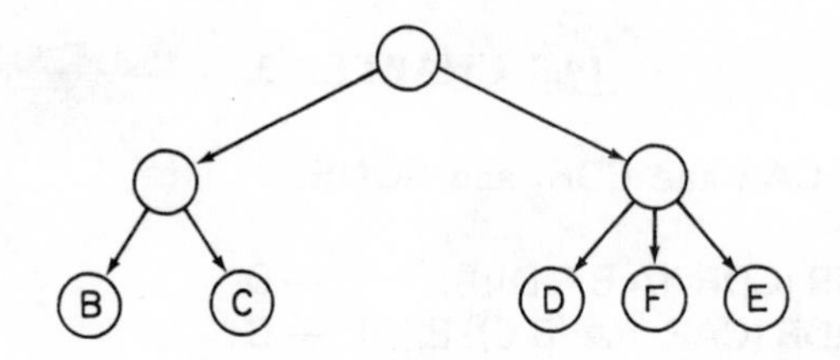

and finally (((A) (B C) (D F E)) (H) G (K L))

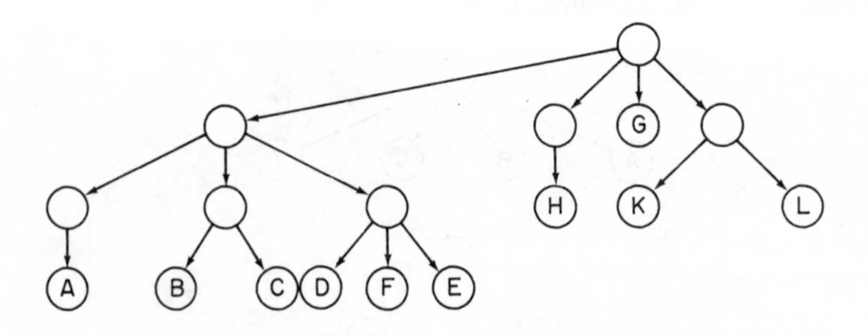

5. Translation of the function calls in the preceding exercise into the form of lists:

```
(CAR '((A) (B C) (D F E)))
(CDR '((A) (B C) (D F E)))
(CONS '((A) (B C) (D F E)) '((H) G (K L)))
```

19.4 CHAPTER 4

1. Results of calls to function QUICK:

```
((A B C) 1 2 3)
((JA ES GEHT) OUI CA VA)
((A CAR) UNE VOITURE)
```

2. Combinations of user functions:

```
(MI MI MI)
((UN BRAVO LA TABLE) (SUR BRAVO BRAVO) CUBE ROUGE)
```

3. A greeting function:

```
(DEFUN BONJOUR ()
  '(BONJOUR))
```

4. Four repetitions of the same element:

```
(DEFUN 4TIMES (ARGUMENT)
       (CONS ARGUMENT
         (CONS ARGUMENT
           (CONS ARGUMENT
             (CONS ARGUMENT ()))))))
```

5. A function which inverts a list of three elements:

```
(DEFUN REVERSE3 (ARG1 ARG2 ARG3)
        (CONS ARG3 (CONS ARG2 (CONS ARG1 NIL))))
```

19.5 CHAPTER 5

19.5.1 Chapter 5.1.1

1. A function which tests if the second element of a list is a number:

```
(DEFUN NUMBERP-CADR (L)
      (NUMBERP (CADR L)))
```

2. A function which tests if the first element of a list is a list:

```
(DEFUN LIST-CAR? (L)
              (CONS 'IS (CONS (CONSP (CAR L)) ())))
```

or:

```
(DEFUN LIST-CAR? (L)
      (CONS 'IS (CONS (NULL (ATOM (CAR L))) NIL)))
```

3. As defined, NUMBERP-CADR tests whether the *first* element of a list is a number. These calls therefore give the results: T, NIL, T, T.

19.5.2 Chapter 5.2.1

1. A function which tests whether the first three elements of a list are numbers:

```
(DEFUN 3NUMBERS (L)
       (IF (NUMBERP (CAR L))
         (IF (NUMBERP (CADR L))
           (IF (NUMBERP (CADDR L)) 'BRAVO 'LOSE)
           'LOSE)
         'LOSE))
```

2. A function which inverts a list of 1, 2 or 3 elements:

```
(DEFUN REV (L)
       (IF (NULL (CDR L)) L
         (IF (NULL(CDR (CDR L)))(CONS (CADR L)(CONS (CAR L) ()))
           (CONS (CADDR L) (CONS (CADR L) (CONS (CAR L) NIL))))))
```

3. The calls to function WEIRD give the following results:

```
(WELL,THEN,WHAT)
(1 2 3)
(1 2 3)
(WELL ITS YOU)
(NUMBER)
(DO IS NOT A LIST, BUDDY)
```

19.6 CHAPTER 6

1. The new DELETE function, which removes any occurrence of its first argument at whatever depth, differs from the version because it contains a recursive call to the CAR of the list (if the first element is a list) and because it uses the auxiliary function PASTE.

   This PASTE function is used if a sub-list contains only elements to eliminate, as in the call

   (DELETE 'A '(1 (A A) 2 3))

   The result should be (1 2 3), i.e. a list from which the whole sub-list has been removed. If we had used function CONS instead of this new function, the result would have been (1 NIL 2 3), which is less elegant.

```
(DEFUN DELETE (ELT LST)
       (IF (NULL LST) ()
         (IF (ATOM LST)
           (IF (EQ ELT EST) () LST)
           (IF (CONSP (CAR LST))
             (PASTE (DELETE ELT (CAR LST)) (DELETE ELT (CDR LST)))
             (IF EQ (CAR LST) ELT) (DELETE ELT (CDR LST))
               (CONS (CAR LST) (DELETE ELT (CDR LST)))))))
```

   with function PASTE defined as follows:

```
(DEFUN PASTE (X LST)
       (IF (NULL X) LST
            (CONS X LST)))
```

2. A function which doubles all the elements in a list:

```
(DEFUN DOUBLE (L)
       (IF (NULL L) ()
        (CONS (CAR L) (CONS (CAR L) (DOUBLE (CDR L))))))
```

3. And here is a function which doubles all the atoms in a list:

```
(DEFUN DOUBLE (L)
       (IF (NULL L) ()
        (IF (CONSP (CAR L))
         (CONS (DOUBLE (CAR L)) (DOUBLE (CDR L)))
           (CONS (CAR L) (CONS (CAR L) (DOUBLE (CDR L)))))))
```

4. Here is function MEMQ which looks for occurrences of an element within a list:

```
(DEFUN MEMQ (ELT LST) (COND
       ((NULL LST) NIL)
       ((EQ (CAR LST) ELT) LST)
       (T (MEMQ ELT (CDR LST)))))
```

5. Function MEMBER differs from function MEMQ as follows

- it searches for elements of any type: the necessary modification is to replace the call to function EQ by a call to function EQUAL
- it searches for an element at any level: as above, with functions DELETE and DOUBLE, the solution will be to apply the function not only to successive CDRs, but also to each CAR, each time the CAR is a list.

Here is the new function:

```
(DEFUN MEMBER (ELT LST) (COND
       ((ATOM LST) NIL)
       ((EQUAL (CAR LST) ELT) LST)
       (T (MEMB1 ELT (MEMBER ELT (CAR LST)) (CDR LST)))))

(DEFUN MEMB1 (ELT AUX LST)
        (IF AUX AUX (MEMBER ELT (CDR LST))))
```

6. A function which groups the successive elements in two lists:

```
(DEFUN GR (L1 L2)
       (IF (NULL L1) L2
        (IF (NULL L2) L1
         (CONS
           (CONS (CAR L1) (CONS (CAR L2) ()))
           (GR (CDR L1) (CDR L2))))))
```

7. Start with function FOOBAR: it builds a list including as many occurrences of it first argument as there are elements in the list specified in the second argument Here are a few examples of calls:

```
(FOOBAR '(X Y Z) '( 1 2 3)) → ((X Y Z)(X Y Z)(X Y Z))
(FOOBAR '(1 2 3) '(X Y Z))  → ((1 2 3)(1 2 3)(1 2 3))
(FOOBAR 1 '(A A A A A A)) → (1 1 1 1 1 1)
```

Next, function FOO: it constructs a list containing the results of the calls t FOOBAR with the list given as argument, and all the successive CDRs of that list

If we submit a list with x elements to FOO, the result list will be a list beginnin with x repetitions of the list, followed by $x - 1$ occurrences of the CDR of tha list, followed by $x - 2$ occurrences of the CDR of the CDR of that list, and s on, until there is a single occurrence of a list consisting of only the last elemen of the list given as argument.

For example, if we make the following call to FOO:

```
(FOO '(X Y Z))
```

the result is:

```
((X Y Z)(X Y Z)(X Y Z)(Y Z)(Y Z)(Z))
```

8. The really exciting thing about function F is that it permutates its argument a each recursive call.

This function builds up a list in which the elements of the first argument an those of the second are mixed, with their order preserved. Here are three calls t this function:

```
(F '(A B C D) '(1 2 3 4)) → (A 1 B 2 C 3 D 4)
(F '(A B C D) '(1 2))     → (A 1 B 2 C D)
(F '(A B C) '(1 2 3 4 5)) → (A 1 B 2 C 3 4 5)
```

9. This surprising function BAR (with four recursive calls) calculates the inverse o the list given as argument. Its effect is therefore identical to that of functio REVERSE with the second argument equal to NIL.

To understand how it works, run it by hand, but do not choose too long a list.

19.7 CHAPTER 7

1. For function NTH, we assume that the numerical argument is greater than o equal to 1, otherwise the function has little sense: what is the '-3rd' element of list?

```
(DEFUN NTH (NUMBER LST) (COND
        ((NULL LST) NIL)
        ((<= N 1) (CAR LST))
        (T (NTH (1- NUMBER) (CDR LST)))))
```

2. Writing a function to transform decimal numbers into octal and hexidecimal numbers is much easier if we divide the task into several sub-tasks: we start by writing the function DEC-OCT which translates decimal numbers into octal, and then function DEC-HEX which translates decimal numbers into hexadecimal, before writing the main function DEC-OCT-HEX which simply makes a list out of the results of calls to the two previous functions.

```
(DEFUN DEC-OCT (N)
   (IF (> N 0)
      (APPEND (DEC-OCT (FLOOR N 8)) (APPEND (REM N 8) NIL)) NIL))

(DEFUN DEC-HEX (N)
   (IF (> N 0)
      (APPEND (TRANS (DEC-HEX (FLOOR N 16)))
               (APPEND (REM N 16) NIL)) NIL))
```

Note that we used the APPEND defined in the text.

Function TRANS is used to translate values between 10 and 15 into their corresponding hexadecimal representations (A to F).

```
(DEFUN TRANS (L)
        (IF (NULL L)()
         (CONS
           (CNTH(1+ (CARL))'(0 1 2 3 4 5 6 7 8 9 A B C D E F)))
           (TRANS (CDR L)))))

(DEFUN DEC-OCT-HEX (NUMBER)
        (CONS (DEC-HEX NUMBER) (CONS (DEC-OCT NUMBER)())))
```

3. Function OCT-DEC converts octal numbers into decimal, while function HEX-DEC converts hexadecimal into decimal.

```
(DEFUN OCT-DEC (N) (OCT-DEC1 N O))

(DEFUN OCT-DEC1 (N RES)
     (IF N (OCT-DEC1 (CDR N) (+ (* RES 8) (CAR N))) RES)))

(DEFUN HEX-DEC (N) (HEX-DEC1 N O))

(DEFUN HEX-DEC1 (N RES)
     (IF N (HEX-DEC1 (CDR N) (+ (* RES 16)
                                 (IF (NUMBERP (CAR N))) (CAR N)
```

```
                           (LET ((M (CAR N)) (COND
                              ((EQ M 'A) 10)
                              ((EQ M 'B) 11)
                              ((EQ M 'C) 12)
                              ((EQ M 'D) 13)
                              ((EQ M 'E) 14)
                              ((EQ M 'F) 15))))))
    RES)))
```

4. And here is the FIBONACCI function:

```
(DEFUN FIBONACCI (N M)
    (IF (< N 1) 1
    ( + (FIBONACCI (1- N)) (FIBONACCI (- N 2)))))
```

Here is a terminal recursive version of the same function:

```
(DEFUN FIBONACCI (N)
       (IF (<= N 1) 1 (FIB 2 '(1 1) N)))

(DEFUN FIB (COUNTER LAST N)
       (IF (= COUNTER N) (+ (CAR LAST) (CADR LAST))
         (FIB (1+ COUNTER)
           (CONS (CADR LAST) (CONS (+ (CAR LAST) (CADR LAST))
                                   NIL))
                                   N)))
```

5. Function WHAT "inverts" a number, i.e. if WHAT is called with a number in the form $c_1 c_2 . . . c_n$, the result will be a number in the form $c_n c_{n-1} . . . c_1$. For example, the result of a call (WHAT 1234) will be the number 4321.

6. Here is Ackermann's function:

```
(DEFUN ACK (M N) (COND
    ((ZEROP M) (1+ N))
    ((ZEROP N) (ACK (1- M) 1))
    (T (ACK (1- M) (ACK M (1- N))))))
```

7. Here are two versions of a function which calculates the length of a list. Only the first version is terminal recursive.

```
(DEFUN LENGTH (L) (LENGTH-AUX L 0))

(DEFUN LENGTH-AUX (L N)
   (IF (NULL L) N
        (LENGTH-AUX (CDR L) (1+ N)))))

   (DEFUN LENGTH (L)
        (IF (NULL L) 0 (1+ (LENGTH (CDR L)))))
```

Function NBATOM is identical to function LENGTH, except that it takes all the sub-lists into account. Here too we give two versions:

```
(DEFUN NBATOM (L) (NB1 L 0))

(DEFUN NB1 (L N)
    (IF (NULL L) N
      (IF (ATOM (CAR L)) (NB1 (CDR L) (1+ N))
        (NB1 (CDR L) (NB1 (CAR L) N))))))

(DEFUN NBATOM (L)
      (IF (NULL L) 0
        (IF (ATOM L) 1
          (+ (NBATOM (CAR L)) (NBATOM (CDR L))))))
```

19.8 CHAPTER 8

1. Clearly, if we want to represent these different relationships, all we have to do is to enter them as indicators in P-lists:
   a. (PUT 'GUY 'FATHER 'PETER)
      or (PUT 'PETER 'CHILD 'GUY))
   b. (PUT 'MARY 'FATHER 'PETER)
      or (PUT 'PETER 'CHILD (CONS 'MARY (GET 'PETER CHILD)))
   c. (PUT 'JAMES 'FATHER 'PETER)
      or (PUT 'PETER 'CHILD (CONS 'JAMES (GET 'PETER CHILD)))
   d. (PUT 'PETER 'SEX 'MALE)
   e. (PUT 'MARY 'SEX 'FEMALE)
   f. (PUT 'GUY 'SEX 'MALE) (PUT 'JAMES 'SEX 'MALE)
   g. (PUT 'JUDY 'SEX (GET 'MARY 'SEX))
   h. (PUT 'JUDY 'AGE 22)
   i. (PUT 'ANNE 'AGE 40)
   j. (PUT 'SARAH 'AGE (+ (GET 'JUDY 'AGE) (GET 'ANNE 'AGE)))

2. This is our first attempt at writing a *real* program: the first exercise in which we are not writing a single function, but analysing the problem, breaking it down into sub-problems, identifying the sub-problems common to different tasks, etc.

We shall not give a complete solution to this exercise, but merely point out the guidelines that should be followed.

Each book is represented by an atom whose name corresponds to the code for the book. So, for example, A1, A2, A3 etc are books. The catalogue for the whole library is given by a function, which we call LIB, defined as follows:

```
(DEFUN LIB () '(A1 A2 A3 . . . AN))
```

Consequently a call to

(LIB)

would return the list:

(A1 A2 A3 . . . AN)

Next, we have to associate properties with each book: a title, an author and a list of keywords. The natural position for these properties is in the P-lists of the book atoms. For example, for Karl Kraus's two books, we would have:

```
(PUT 'A1 'AUTHOR '(KARL KRAUS))
(PUT 'A1 'TITLE '(DIE LETZTEN TAGE DER MENSCHHEIT))
(PUT 'A2 'AUTHOR '(KARL KRAUS))
(PUT 'A2 'TITLE '(VON PEST UND PRESSE))
```

What we have to do now, after having updated the library, is to write a set of functions allowing it to be consulted.

Let us first build up an auxiliary function which searches through the library for all occurrences of a certain property:

```
(DEFUN SEARCHALL (PROP VAL) (SEARCH PROP (LIB) VAL))

(DEFUN SEARCH (PROP BASE VAL) (COND
          ((NULL BASE) NIL)
          ((EQUAL (GET (CAR BASE) PROP) VAL)
                 (CONS (CAR BASE) (SEARCH PROP (CDR BASE) VAL)))
          ((MEMBER VAL (GET (CAR BASE) PROP))
                 (CONS (CAR BASE) (SEARCH PROP (CDR BASE) VAL)))
          (T (SEARCH PROP (CDR BASE) VAL))))
```

This function makes it possible, for example, to find all books by a particular author. For example, if books A1, A2 and A23 are by Karl Kraus and book A456 is by Karl Kraus and Ludwig Wittgenstein, the call

(SEARCHALL 'AUTHOR '(KARL KRAUS)

would give us the list (A1 A2 A23 A456).

We also need a function, which we might call FORALL, which will give us the property required for each of these books. Here it is:

```
(DEFUN FORALL (LST PROPERTY)
    (IF (NULL LST) NIL
        (LET ((AUX (GET (CAR LST) PROPERTY)))
          (IF AUX (CONS AUX (FORALL (CDR LST) PROPERTY))
              (FORALL (CDR LST) PROPERTY)))))
```

We can now start writing query functions for the database. First we shall write a function making it possible to find the name of an author given the title of a book:

```
(DEFUN AUTHOR (TITLE)
     (FORALL (SEARCHALL 'TITLE TITLE) 'AUTHOR))
```

Note that this function will also handle cases where two authors have written books with the same title.

Next we have a function which will give us all the books by an author:

```
(DEFUN ALLTITLE (AUTHOR)
       (FORALL (SEARCHALL 'AUTHOR AUTHOR) 'TITLE))
```

To end with, here is a function which will give us all the books referring to given subjects in the list of keywords, which is held in property KEYWORD.

```
(DEFUN ALLBOOK (KEYWORD)
     (FORALL (SEARCHALL 'KEYWORD KEYWORD) 'TITLE))
```

To find all the authors who have written on a particular subject, we simply replace property TITLE by property AUTHOR, in the last function.

That is as far as well shall go in developing our (simple) bibliographic database query language. We leave it up to you to go further.

. Start by assigning the right properties to the various atoms:

```
(PUT 'ALBERT 'MOTHER 'MARTINE)   (PUT 'ALBERT 'FATHER 'CLAUD)
(PUT 'PETER 'MOTHER 'ELIZABETH)  (PUT 'PETER 'FATHER 'JOHN)
(PUT 'COLIN 'MOTHER 'EVE)        (PUT 'COLIN 'FATHER 'ALBERT)
(PUT 'GERALD 'MOTHER 'JULIE)     (PUT 'GERALD 'FATHER 'PETER)
(PUT 'GUY 'MOTHER 'JULIE)        (PUT 'GUY 'FATHER 'PETER)
(PUT 'MARY 'MOTHER 'JULIE)       (PUT 'MARY 'FATHER 'PETER)
(PUT 'JAMES 'MOTHER 'JULIE)      (PUT 'JAMES 'FATHER 'PETER)
(PUT 'ALAN 'MOTHER 'ALICE)       (PUT 'ALAN 'FATHER 'COLIN)
(PUT 'BRIAN 'MOTHER 'KATHERINE)  (PUT 'BRIAN 'PETER 'ALAN)
(PUT 'ANNE 'MOTHER 'SARAH)       (PUT 'ANNE 'PETER 'GERALD)
(PUT 'KATHERINE 'MOTHER 'JANE)   (PUT 'KATHERINE 'FATHER 'JAMES)
(PUT 'JUDY 'MOTHER 'ANNE)        (PUT 'JUDY 'FATHER 'PAUL)
```

etc.
Now for their sexes:

```
(DEFUN PUT-SEX (SEX LST)
     (IF (NULL LST) NIL
        (PUT (CAR LST) 'SEX SEX)
        (PUT-SEX SEX (CDR LST))))
```

```
(PUT-SEX 'MALE)
  '(CLAUDE JOHN ALBERT PETER COLIN GERALD GUY JAMES ALAN
  BRIAN PAUL))

(PUT-SEX 'FEMALE)
     '(ELIZABETH EVE JULIE ALICE SARAH MARY JANE ANNE KATHERINE
     JUDY))
```

As in the preceding exercise, we shall now construct a function which calculates a list of all the people:

```
(DEFUN PEOPLE ()
      ' (CLAUDE ELIZABETH JOHN ALBERT EVE PETER JULIE ALICE
      COLIN SARAH GERALD GUY MARY JAMES JANE ALAN ANN
      KATHERINE PAUL BRIAN JUDY))
```

Next, we only need to write functions which deduce family relations from this database. Let us start with a function which finds brothers:

```
(DEFUN BROTHERS (X)
      (LET ((FATHER (GET X 'FATHER)) (MOTHER (GET X 'MOTHER)))
           (FINDALL 'MOTHER MOTHER
                (FINDALL 'FATHER FATHER
                  (FINDALL 'SEX 'MALE (PEOPLE))))))
```

Naturally, function FINDALL does not yet exist. What form should it take? Look carefully at the problem: knowing that brothers must all be male, that they must all have the same father and mother, I took the decision that in order to find all somebody's brothers, I had to look through all people known to be male, for all those who have the same father, and then look through this new set of people for those who have the same mother.

This means that I shall have to go through lists three times looking for elements with the same property in common. Instead of writing the same thing three times, I felt it was best to write an auxiliary function to carry out this task. Here is the little function FINDALL:

```
(DEFUN FINDALL (INDICATOR VALUE LST) (COND
       ((NULL LST) ())
       ((EQUAL (GET (CAR LST) INDICATOR) VALUE)
            (CONS (CAR LST) (FINDALL INDICATOR VALUE (CDR LST))))
       (T (FINDALL INDICATOR VALUE (CDR LST)))))
```

Look at this function: it contains two recursive calls. In these two calls, only *one* argument, LST, is modified. The other two arguments remain constant. Knowing that each call to a function implies a new bindings of parameter variables, we can slightly improve this function by taking the invariant argument out of the loop.

Here then is a second version of function FINDALL, first for Le_LISP.

```
(DE FINDALL (INDICATOR VALUE LST)
      (LETN SELF ((LST LST)) (COND
            ((NULL LST) ())
            ((EQUAL (GET (CAR LST) INDICATOR) VALUE)
                (CONS (CAR LST) (SELF (CDR LST))))
            (T (SELF (CDR LST))))))
```

and then for Common-LISP:

```
(DEFUN FINDALL (INDICATOR VALUE LST)
      (LABELS ((SELF (LST) (COND
                  ((NULL LST) NIL)
                  ((EQUAL (GET (CAR LST) INDICATOR) VALUE)
                          (CONS (CAR LST) (SELF (CDR LST))))
                  (T (SELF (CDR LST))))))
        (SELF LST)))
```

But let us continue. A function which finds a person's sisters will not be difficult to write: all we have to do is to replace the property MALE by the property FEMALE, in function BROTHERS. This gives us:

```
(DEFUN SISTERS (X)
      (LET ((FATHER (GET X 'FATHER))(MOTHER (GET X 'MOTHER)))
          (FINDALL 'MOTHER MOTHER
              (FINDALL 'FATHER FATHER
                  (FINDALL 'SEX 'FEMALE (PEOPLE))))))
```

Now we have two almost identical functions: BROTHERS and SISTERS. Every time you find almost identical functions in a program, you can be sure that a better way of doing things is to construct an auxiliary function which contains the common points. Let us therefore write an auxiliary function which we shall call FINDALL-THREE, as it has to find all elements which have three identical properties:

```
(DEFUN FINDALL-THREE (X PROP1 PROP2 PROP3 VALUE3)
      (FINDALL PROP1 (GET X PROP1)
        (FINDALL PROP2 (GET X PROP2)
          (FINDALL PROP3 VALUE3 (PEOPLE)))))
```

and functions BROTHERS and SISTERS can be simplified to:

```
(DEFUN BROTHERS (X)
      (DELETE X (FINDALL-THREE X 'FATHER 'MOTHER 'SEX 'MALE)))

(DEFUN SISTERS (X)
      (DELETE X (FINDALL-THREE X 'FATHER 'MOTHER 'SEX 'FEMALE)))
```

and to obtain brothers and sisters at the same time, we simply have to write:

```
(DEFUN SIBLINGS (X) (DELETE X (APPEND (BROTHERS X) (SISTERS X))))
```

We shall leave it up to you to complete the exercise by writing the necessary functions to find cousins, male or female, aunts, uncles, etc. All these functions should now be easy to write.

To conclude this exercise, here is the function which gives you all the ancestors of a particular person:

```
(DEFUN ANCESTORS (X)
       (LET ((MOTHER (GET X 'MOTHER))(FATHER (GET X 'FATHER)))
         (IF (NULL MOTHER)
           (IF (NULL FATHER) () (CONS FATHER (ANCESTORS FATHER)))
           (IF (NULL FATHER)
            (CONS MOTHER (ANCESTORS MOTHER))
            (CONS MOTHER (CONS FATHER
              (APPEND (ANCESTORS FATHER)
                (ANCESTORS MOTHER)))))))))
```

19.9 CHAPTER 9

1. Here is function FACTORIAL as a memo function:

```
(DEFUN FACTORIAL (N) (COND
         ((<= N 0) 1)
         ((GET 'FACTORIAL N))
         (T (PUT 'FACTORIAL N (*N (FACTORIAL (1- N))))
           (GET 'FACTORIAL N))))
```

2. Given that a COND evaluates one test after another, we can rewrite this program as follows:

```
(DEFUN FOO (X) (COND
       ((< X 0) (- 0 X)
       ((ZEROP X) 0)
       (T (* X -1))))
```

Clearly, this program could also be simplified to read:

```
(DEFUN INVERSE (X) (- 0 X))
```

3. Here is a first algebraic simplifier:

```
(DEFUN SIMP (X)(COND
       ((ATOM X) X)
```

```
((EQ (CAR X) '+) (SIMPLIFY-PLUS (SIMP (CADR X))(SIMP (CADDR X))))
((EQ (CAR X) '*)
  (SIMPLIFY-TIMES (SIMP (CADR X)) (SIMP (CADDR X))))
(T X)))

(DEFUN SIMPLIFY-PLUS (X Y) (COND
       ((AND (NUMBERP X) (NUMBERP Y)) (+ X Y))
       ((ZEROP X) Y)
       ((ZEROP Y) X)
       (T (CONS '+ (CONS X (CONS Y ())))))))

(DEFUN SIMPLIFY-TIMES (X Y) (COND
       ((AND (NUMBERP X) (NUMBERP Y)) (* X Y))
       ((OR (ZEROP X) (ZEROP Y)) 0)
       ((= X 1) Y)
       ((= Y 1) X)
       (T (CONS '* (CONS X (CONS Y ())))))))
```

In this program we use two new functions, AND and OR.

Function AND is the logical *and*. It takes any number of arguments and if *all* the arguments have evaluated to values other than NIL, it returns the value of the last argument; otherwise, it returns NIL.

Function OR is the logical *or*. It also takes any number of arguments and returns the value of the *first* argument which evaluates to a value different from NIL. If *all* the arguments evaluate to NIL, the OR function returns NIL.

Clearly, we could have written the program without these two new standard functions. But writing it with AND and OR is more readable. See for yourself: here is SIMPLIFY-TIMES without these new functions:

```
(DEFUN SIMPLIFY-TIMES (X Y)(COND
       ((ZEROP X) 0)
       ((ZEROP Y) 0)
       ((= X 1) Y)
       ((= Y 1) X)
       ((NUMBERP X)
         (IF (NUMBERP Y)(* X Y)
           (CONS '* (CONS X (CONS Y ())))))))
       (T (CONS '* (CONS X (CONS Y ()))))))
```

19.10 CHAPTER 10

1. To generalize this little program so that it can conjugate verbs in other tenses as well as the present, the first thing to resolve is how to represent the additional endings.

As things now stand, our endings are in the middle of function CONJ1. Holding data in the middle of code is very seldom a good methodology. It is often preferable (and more readable) to separate data from algorithms. This is what we shall do by creating new atoms ER and IR. We shall put the lists of endings, under indicators corresponding to the tenses, in the P-lists of these two atoms. This gives us:

```
(PUT 'IR
            'present
              '(is is it it issons issez issent issent)
            'imperfect
              '(issais issais issait issait issions issiez issaient issaient)
            'past historic
              '(is is it it imes ites irent irent)
            'future
              '(irai iras ira ira irons irez iront iront)
            'conditional
              '(irais irais irait irait irions iriez iraient iraient)
            'present-subjunctive
              '(isse isses isse isse issions issiez issent issent)
            'imperfect-subjunctive
              '(isse isse it it issions issiez issent issent))

(PUT 'ER
            'present)
              '(e es e e ons ez ent ent)
            'imperfect
              '(ais ais ait ait ions iez aient aient)
            'past-historic
              '(ai as a a ames ates erent erent)
            'future
              '(erai eras era era erons erez eront eront)
            'conditional
              '(erais erais erait erait erions eriez eraient eraient)
            'present-subjunctive
              '(e es e e ions iez ent ent)
            'imperfect-subjunctive
              '(asse asses at at assions assiez assent assent))
```

Now we have to change function CONJ1 so as to remove the lists of endings from its code and to adapt the program to these modifications. We have to replace the lines:

```
(LET ((ENDINGS (COND
  ((EQ TYPE 'ER) '(E ES E E ONS EZ ENT ENT))
  ((EQ TYPE 'RE) '(S S T T ONS EZ ENT ENT))
  (T '(IS IS IT IT ISSONS ISSEZ ISSENT ISSENT)))))
```

with the single line:

```
(LET ((ENDINGS (GET TYPE TENSE)))
```

assuming that variable TENSE has the value of the conjugation requested. TENSE therefore becomes the second argument of function CONJUGATE:

```
(DEFUN CONJUGATE (VERB TENSE)...
```

Next we replace the call to function PRINT, within CONJ1, by a call to function VERBPRINT as follows:

```
(DEFUN VERBPRINT (PRONOUN VERB TENSE)
    (LET ((AUX (MEMQ TENSE '(present-subjunctive imperfect-subjunc-
    tive))))
      (FORMATT "~Z~A~A ~A"
        (IF AUX
          (IF (MEMQ PRONOUN '(il elle ils elles)) "qu'" "que ") "")
        PRONOUN VERB)))
```

and that should do the trick.

However, as you will doubtless have noticed, nothing has yet been provided for conjugating composite tenses. To handle them, the system has to know the necessary forms of the verb *avoir*. Let us include them on the P-list for AVOIR:

```
(PUT 'AVOIR
  'perfect
    '(ai as a a avons avez ont ont)
  'pluperfect
    '(avais avais avait avions aviez avaient avaient)
  'past-anterior
    '(eus eus eut eut eumes eutes eurent eurent)
  'future-anterior
    '(aurai auras aura aura aurons aurez auront auront)
  'past-conditional1
    '(aurais aurais aurait aurait aurions auriez auraient auraient)
  'past-conditional2
    '(eusse eusses eut eut eussions eussiez eussent eussent))
```

Now we have to build a special print function for composite tenses, as we did for the subjunctive:

```
(DEFUN PR-COMPOS (PRON PARTICIPLE TENSE ROOT)
      (LABELS ((DO-IT (AUX PRON PARTICIPLE) (COND
                        ((NULL AUX) "that's it")
                        (T (FORMATT "~%~A ~A ~A"
```

```
                    (IF (EQ (CAR PRON) 'je) "j'" (CAR PRON))
                    (CAR AUX) PARTICIPLE)
                   (DO-IT (CDR AUX) (CDR PRON) PARTICIPLE))))
      (DO-IT (GET 'AVOIR TENSE)
             PRON
             (CREATE-WORD ROOT PARTICIPLE))))
```

All that remains is to change function CONJ1 to allow for the construction c composite tenses. Remember that for the moment we have a function which ca conjugate normally (i.e. non-composite tenses) and another which will allo composite tenses. We therefore have to introduce a test which will decide whic is to be called.

Function CONJ1 below resolves the problem by remaining inside CONJ1 if th evaluation of (GET TYPE TENSE) differs from NIL, and to launch function PR COMPOS otherwise, i.e. if the P-list of TYPE (i.e. ER or IR), does not contai a list of endings associated with the tense requested:

```
(DEFUN CONJ1 (ROOT TYPE PRONOUNS TENSE)
   (LABELS ((DO-IT (PRONOUNS ENDINGS) (COND
                 ((NULL PRONOUNS) "that's it")
                 ((ATOM ENDINGS)
                       (PR-COMPOS PRONOUNS
                                  (IF (EQ TYPE 'IR) 'I 'E)
                                  TENSE ROOT))
                 (T (VERBPRINT (CAR PRONOUNS)
                               (CREATE-WORD ROOT (CAR ENDINGS)
                               TENSE)
                    (DO-IT (CDR PRONOUNS) (CDR ENDINGS))))))
      (DO-IT PRONOUNS (GET TYPE TENSE))))
```

this makes it necessary to adapt the call to CONJ1. So, here is the new versio of CONJUGATE:

```
(DEFUN CONJUGATE (VERB TENSE)
      (TERPRI)
      (CONJ1 (ROOT VERB) (TYPE VERB)
            '(JE TU IL ELLE NOUS VOUS ILS ELLES) TENSE))
```

2. This exercise we shall leave entirely to you. After the previous exercise it shoul not in any case give you any problems.

3. It must have become clear that problems can be solved (and therefore programmed in many different ways. Here is *one* way of solving the problem of the Hano Towers with some disks on each needle at the start:

```
(DEFUN HANOI (N S F I &OPTIONAL AID)
   (WHEN (> N 0)
```

```
        (IF AID (HANOI (1- N) S I F) (HANOI N I F S N))
        (FORMAT T "~%disk ~A from ~A to ~A" N S F)
        (IF (AND AID (= N AID)) (HAN (1- N) I S F) (HAN (1- N) I F S))))

(DEFUN HAN (N S F I)
  (LABELS ((PRINT-HANOI (C)
                (WHEN (> C 0)
                      (FORMAT T "~%disk ~A from ~A to ~A" N S F)
                      (PRINT-HANOI (1- C)))))
       (WHEN (> N 0)
             (HAN (1- N) S I F)
             (PRINT-HANOI 3)
             (HAN (1- N) I F S))))
```

Study this function carefully: it looks difficult to understand.

Note that I used function WHEN. This function is similar to function IF: the first argument is a *test* and if this test evaluates to something other than NIL the sequence of instructions forming the remainder of the WHEN function's arguments is executed and the value of the last expression is that of the WHEN; otherwise, i.e. if the test evaluates to NIL, we leave the WHEN and return value NIL.

Here are two calls to this function (if you want to see the result with more disks, implement, improve, and test the program yourselves):

```
?    (HANOI 1 START 'FINAL 'INTERMEDIATE)
     disk 1 from INTERMEDIATE to FINAL
     disk 1 from START to FINAL
=    NIL

?    (HANOI 2 'START 'FINAL 'INTERMEDIATE)
     disk 1 from FINAL to START
     disk 1 from INTERMEDIATE to START
     disk 2 from INTERMEDIATE TO FINAL
     disk 1 from START to INTERMEDIATE
     disk 1 from START to INTERMEDIATE
     disk 1 from START to INTERMEDIATE
     disk 2 from START to FINAL
     disk 1 from INTERMEDIATE to START
     disk 1 from INTERMEDIATE to START
     disk 1 from INTERMEDIATE to START
=    NIL
```

19.11 CHAPTER 11

1. Function TIMES, which multiplies a sequence of numbers, can be defined as follows:

   (DE TIMES L (TTIMES L))

```
(DE TTIMES (L)
  (IF (NULL (CDR L)) (CAR L)
    (* (CAR L) (TTIMES (CDR L)))))
```

2. Here is the function which constructs a *list of values* of elements of even rank:

```
(DE SLIST L (SSLIST (CDR L)))

(DE SSLIST (L)
  (IF (NULL L) ()
    (CONS (CAR L) (SSLIST (CDDR L)))))
```

3. Here is the function which constructs a *list of elements* of even rank:

```
(DF QSLIST L (SSLIST (CDR L)))

(DE SSLIST (L)
  (IF (NULL L) ()
    (CONS (CAR L) (SSLIST (CDDR L)))))
```

Clearly, the two SSLIST functions are identical. The only difference is at the leve of definition of function SLIST, which is an NEXPR, and QSLIST, which is a FEXPR.

4. Here is function MAX which finds the greatest of a series of numbers:

```
(DE MAX L (IF L (MAXIMUM (CDR L) (CAR L))))
```

The auxiliary function MAXIMUM takes two arguments: a list of numbers an the maximum found so far. Obviously, there is nothing stopping me taking th first argument as a temporary maximum, and then looking through the remainin list for a larger number, and so forth. Here is the function:

```
(DE MAXIMUM (L N)
  (IF (NULL L) N
        (MAXIMUM (CDR L) (IF (< N (CAR L)) (CAR L) N))))
```

Function MIN will obviously be very similar: the only difference is that instea of the test (> N (CAR L)) we shall have (> N (CAR L)). Here are the two function necessary:

```
(DE MIN L (IF L (MINIMUM (CDR L) (CAR L))))

(DE MINIMUM (L N)
  (IF (NULL L) N
    (MINIMUM (CDR L) (IF (> N (CAR L)) (CAR L) N))))
```

19.12 CHAPTER 12

1. The new version of function TIMES is:

```
(DEFUN TIMES (&REST L)
       (IF (NULL L) 1
         (* (CAR L) (APPLY 'TIMES (CDR L)))))
```

2. The new version of function SLIST is:

```
(DEFUN SLIST (&REST L)
       (IF (NULL (CDR L)) ()
              (CONS (CADR L) (APPLY 'SLIST (CDDR L)))))
```

3. QSLIST is still only possible in Le_LISP. Here is its new version:

```
(DF QSLIST L
  (IF (NULL (CDR L)) ()
   (CONS (CADR L) (EVAL (CONS 'QSLIST (CDDR L))))))
```

Here we cannot use function APPLY, which will not accept an FEXPR type function in the functional position.

4. For our little word by word translator to work at least for the few examples given, we shall need the following vocabulary:

```
(DEFUN CHAT () 'CAT)
(DEFUN LE () 'THE)
(DEFUN MANGE () 'EATS)
(DEFUN SOURIS () 'MOUSE)
(DEFUN A () 'HAS)
(DEFUN VOLE () 'STOLEN)
(DEFUN FROMAGE () 'CHEESE)
```

Then we need a function which will take any sequence of words (which means it will be an FEXPR), converts each of those words into a function call and returns as value the sequence of results of each of these word-functions. Here it is (in Le_LISP):

```
(DF TRANSLATE SENTENCE
  (IF (NULL SENTENCE) ()
   (CONS
    (APPLY (CAR SENTENCE) ()) ;translate a word;
    (EVAL (CONS 'TRANSLATE   ;translate the rest;
       (CDR SENTENCE))))))   ;of the sentence;
```

19.13 CHAPTER 13

1. Here is the macro-character '[circumflex]' which allows us to load a file simply by writing:

   ^name-of-file

```
(SET-MACRO-CHARACTER #\^ '(LAMBDA (STREAM CHAR)
        ['LOAD (READ STREAM T NIL T)]))
```

   Note that once you have defined a macro-character, you can use it in the definition of other macro-characters. Here, for example, we have used the macro-characters '[' and ']' defined in this chapter.

2. The difficulty of writing these two macro-characters is that they are made up of more than one character. The test that they are followed by the character '=' has therefore to be carried out within the macro-characters '>' and '<'. Here are the definitions of these two macro-characters:

```
(SET-MACRO-CHARACTER #\> '(LAMBDA (STREAM CHAR)
   (LET ((X (PEEK-CHAR STREAM T NIL T)))
        (COND ((EQUAL X #\=) (READ-CHAR)
                  (LIST '|>=| (READ STREAM T NIL T)
                              (READ STREAM T NIL T)))
              (T (LIST '|>| (READ STREAM T NIL T)
                            (READ STREAM T NIL T)))))))

(SET-MACRO-CHARACTER #\< '(LAMBDA (STREAM CHAR)
   (LET ((X (PEEK-CHAR STREAM T NIL T)))
        (COND ((EQUAL X #\=) (READ-CHAR)
              (LIST '|<=| (READ STREAM T NIL T)
                          (READ STREAM T NIL T)))
              (T (LIST '|<| (READ STREAM T NIL T)
                            (READ STREAM T NIL T)))))))
```

3. We will first give a solution in Le_LISP, then in Common-LISP. Here is the Le_LISP solution.

   First we give the program to write even numbers from 2 to N in file even.no:

```
(DE EVEN (N)
      (OUTPUT "even.no")
      (LETN SELF ((X 2)) (COND
        ((<= X N) (PRINT X) (SELF (+ X 2)))
        (T (PRINT "END") (OUTPUT) N))))
```

   Now the same thing for the odd numbers:

```
(DE ODD (N)
       (OUTPUT "odd.no")
       (LETN SELF ((X 1)) (COND
          ((< = X N) (PRINT X) (SELF (+ X 2)))
          (T (PRINT "END") (OUTPUT) N))))
```

Finally, all we have to do is calculate the sum:

```
(DE SUM ()
       (INPUT "odd.no")
       (LETN SELF ((X (READ)) (ODD))
          (IF (EQUAL X "END") (SUM1 (REVERSE ODD))
          (SELF (READ) (CONS X ODD)))))

(DE SUM1 (ODD)
       (INPUT "odd.no")
       (LETN SELF ((X READ)) (EVEN))
          (IF (NULL (EQUAL X "END")) (SELF (READ) (CONS X EVEN))
               (INPUT)
               (SUM2 ODD (REVERSE EVEN)))))

(DE SUM2 (ODD EVEN)
       (OUTPUT "sum.no")
       (LETN SELF ((ODD ODD) (EVEN EVEN))
       (COND
         ((AND (NULL ODD) (NULL EVEN)) (OUTPUT))
         ((NULL ODD) (PRINT (CAR EVEN) (SELF NIL (CDR EVEN)))
         ((NULL EVEN) (PRINT (CAR ODD)) (SELF (CDR ODD) NIL))
         (T (PRINT (+(CAR ODD) (CAR EVEN)))
           (SELF (CDR ODD) (CDR EVEN))))))
```

And here is the same in Common-LISP:

```
(DEFUN EVEN (N)
    (LABELS ((INTERNAL-EVEN (X STREAM)(COND
                  ((<= X N (PRINT X STREAM)
                              (INTERNAL-EVEN (+ X 2) STREAM))
                  (T (PRINT "END" STREAM)(CLOSE STREAM)))))
          (INTERNAL-EVEN 2 (OPEN "even.no" :DIRECTION :OUTPUT))))

(DEFUN ODD (N)
    (LABELS ((INTERNAL-ODD (X STREAM)(COND
                  ((<= X N) (PRINT X STREAM)
                              (INTERNAL-ODD (+ X 2) STREAM))
                  (T (PRINT "END" STREAM)(CLOSE STREAM)))))
          (INTERNAL-ODD 1 (OPEN "odd.no" :DIRECTION :OUTPUT))))
```

```
(DEFUN SUM (ODD EVEN)
   (LABELS ((INTERNAL-SUM (ODD EVEN OUT X Y)(COND
                        ((EQUAL X "END") (CLOSE ODD) (CLOSE EVEN
                                                  (CLOSE OUT))
                        ((EQUAL Y "END") (CLOSE ODD) (CLOSE EVEN
                                                  (CLOSE OUT))
                        (T (PRINT (+ X Y) OUT)
                           (INTERNAL-SUM ODD EVEN OUT (READ EVEN
                                                      (READ ODD))))))
         (LET ((ODD (OPEN ODD :DIRECTION :INPUT))
               (EVEN (OPEN EVEN :DIRECTION :INPUT))
               (OUT (OPEN "sum.no" :DIRECTION :OUTPUT)))
           (INTERNAL-SUM ODD EVEN OUT (READ EVEN)(READ ODD))))
```

19.14 CHAPTER 14

1. From what we have already learnt, all recursive calls which are not the argument of another function, i.e. whose result is not used by another function, are termina recursive calls. Others are normal recursive calls. Below, we again give functio MATCH with terminal recursive calls written in *italics*:

```
(DEFUN MATCH (PATTERN DATA) (COND
           ((ATOM PATTERN)(EQUAL PATTERN DATA))
           ((EQUAL (CAR PATTERN) '&) (COND
                 ((NULL (CDR PATTERN)) T)
                 ((MATCH (CDR PATTERN) DATA))
                 ((DATA (MATCH PATTERN (CDR DATA)))))
                 ((ATOM DATA) NIL)
                 ((EQUAL (CAR PATTERN) '$)
                        (MATCH (CDR PATTERN) (CDR DATA)))
                 ((MATCH (CAR PATTERN) (CAR DATA))
                        (MATCH (CDR PATTERN) (CDR DATA)))))
```

2. To allow the use of the % sign as well, we have to introduce a further clause int function MATCH. Clearly, that clause must test whether the pattern begins wit a list and if that list begins with the special "%" sign:

```
((AND (CONSP (CAR PATTERN)) (EQ (CAAR PATTERN) '|%|)). . .)
```

Next, in the same way we handled matching for *sequences*, we now call MATCH recursively testing each element in turn from the list of candidates. Mor specifically, we shall construct new *patterns*, with a particular pattern for eac candidate, then check whether the pattern corresponds to the first element of th data; if so, we have found an item meeting our requirements, and we have onl to match the rest of the pattern with the rest of the data; if not, we have to tes the pattern formed against the next candidate. Only if none of the patterns built up in this way corresponds to the first element of the data do we have a failure.

Here is the whole clause and the function it calls. This function CHOICE will be a *labelled* function of MATCH.

```
((AND (CONSP (CAR PATTERN))(EQUAL (CAAR PATTERN) '|%|))
     (LET ((AUX (CDR (CAR PATTERN))))
          (CHOICE AUX PATTERN DATA)))

(LABELS ((CHOICE (AUX PATTERN DATA) (COND
                ((NULL AUX) NIL)
                ((MATCH (CAR AUX) (CAR DATA))
                         (MATCH (CDR PATTERN)(CDR DATA)))
                (T (CHOICE (CDR AUX) PATTERN DATA)))))
```

19.15 CHAPTER 15

. The following diagrams show the original lists in solid lines, and modifications in dotted lines.

a. (SET 'X '(A B C)) (ATTACH 1 X)

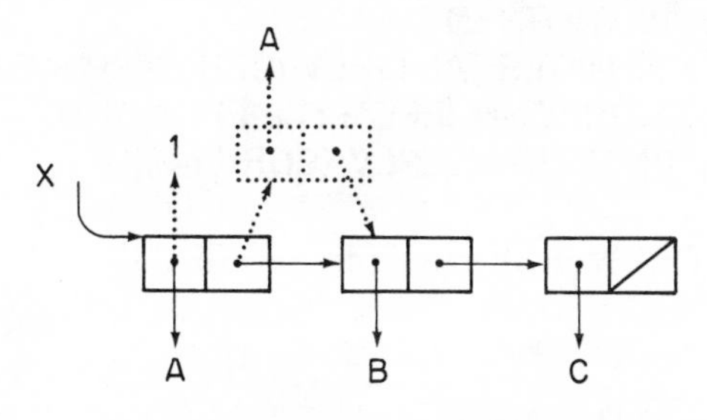

b. (LET ((X '(A B C))) (SMASH X))

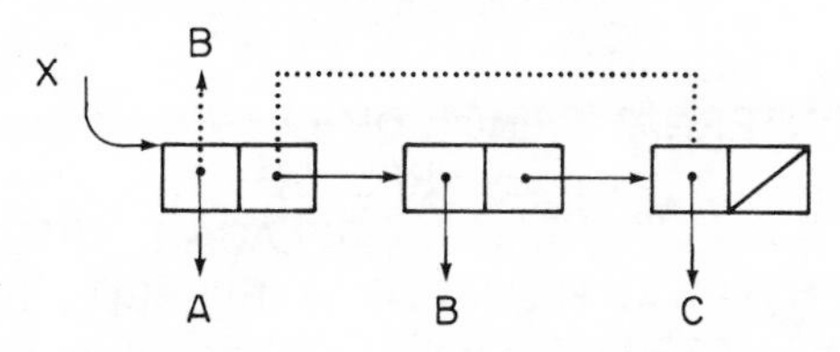

c. (SETQ X '(A B C)) (RPLACA (RPLACD X X) X)

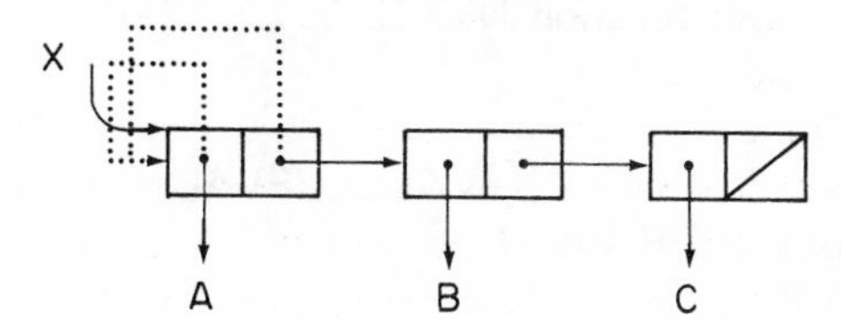

2. Here is a function which physically removes all occurrences of a given elemen from within a list. Function FDELQ is only used in the particular case where th list begins with the element required. If this is the case, the contents of the fir ordered pair have to be replaced by those of the next. In this way, we ensure th all pointers to the original list will subsequently point to the modified list.

```
(DEFUN FDELQ (X Y)(COND
        ((NULL Y) ())
        ((EQUAL X (CAR Y)) (IF (CDR Y) (FDELQ X (SMASH Y)) ()))
        (T (DEL1 X Y))))

(DEFUN DEL1 (X Y) (COND
       ((NULL Y) ())
       ((EQUAL X (CAR Y)) (DEL1 X (CDR Y)))
       ((CONSP (CAR Y))
         (RPLACA Y (DEL1 X (CAR Y)))(RPLACD Y (DEL1 X (CDR Y))))
       (T (RPLACD Y (DEL1 X (CDR Y)))))))
```

3. Clearly, to invert a simple circular list, we have to replace the end of list test b a test for equality between the list pointed to currently and the original list:

```
(DEFUN C-REVERSE (L)
   (LABELS ((INTERNAL-C-REV (C P)
                (IF (NOT (EQ L C))
                   (INTERNAL-C-REV (CDR C)(RPLACD C P))
                   (PROGN (RPLACD L P) P))))
        (AND L (INTERNAL-C-REV (CDR L) L))))
```

4. Finally, function FFREVERSE physically inverts a list (without circularity) at a levels:

```
(DEFUN FREVERSE (L)
   (LABELS ((INTERNAL-FREV (L P)
        (IF L
           (IF (CONSP (CAR L))
              (INTERNAL-FREV (CDR L)
                             (RPLACD
                                (RPLACA L (FREVERSE (CAR L))) P))
              (INTERNAL-FREV (CDR L) (RPLACD L P))) P)))
        (INTERNAL-FREV L NIL)))
```

5. Here is the definition of function NEXTL in Le_LISP

```
(DF NEXTL (VAR)
  (LET ((VARIABLE (CAR VAR)) (VALUE (EVAL (CAR VAR))))
    (SET VARIABLE (CDR VALUE))
    (CAR VALUE)))
```

6. Function RPLACB can be defined as follows:

```
(DEFUN RPLACB (OBJECT LST)
      (RPLACA (RPLACD OBJECT (CDR LST)) (CAR LST)))
```

This function is called DISPLACE in Le_LIST.

19.16 CHAPTER 16

1. Clearly, the macro-functions INCR and DECR are very similar. Here they are:

```
(DEFMACRO INCR (L) '(SETQ ,L (1+ ,L)))
(DEFMACRO DECR (L) '(SETQ ,L (1-, ,L)))
```

2. And here is function SETNTH:
   in LE_LISP:

```
(DEFMACRO SETNTH (WHERE WHICH WHAT)
      '(RPLACA (NTHCDR (1- ,WHICH) ,WHERE) ,WHAT))
```

   and in Common-LISP:

```
(DEFMACRO SETNTH (WHERE WHICH WHAT)
      '(SETF (NTH (1- ,WHICH) ,WHERE) ,WHAT))
```

19.17 CHAPTER 17

1. Here is function MAPS which applies a function to all sub-structures of a list:

```
(DEFUN MAPS (L F) ( COND
        ((NULL L) ())
        ((ATOM L) (F L))
        (T (F L) (MAPS (CAR L) F) (MAPS (CDR L)F)))
```

2. Clearly, the only difference between function WHILE and function UNTIL is the direction of the test: all we have to do is invert it. This gives us:
   in Common-LISP:

```
(DEFMACRO UNTIL (&WHOLE ALL &REST CALL)
         '(IF ,(CAR CALL) ()
                 (PROGN ,@(CDR CALL) ,ALL)))))
```

   in Le_LISP:

```
(DMD UNTIL CALL
          '(IF ,(CADR CALL) NIL
               ,@(CDDR CALL) ,CALL))
```

3. Function FACTORIAL using DO loop would be as follows:

```
(DEFUN FACTORIAL (N)
          (DO ((N N (1- N))
                (R 1 (* N R )))
               ((<= N 1) R)))
```

20 BIBLIOGRAPHY

Abelson 85] H. Abelson and G. J. Sussman, *Structure and Interpretation of Computer Programs*, The MIT Press, Cambridge, Ma (1985).

Allen 78] J. Allen, *The Anatomy of LISP*, McGraw-Hill Inc (1978).

Berkeley 64] E. C. Berkeley and D. G. Bobrow, *The Programming Language LISP; Its Operation and Applications*, Information International Incorporated, Cambridge, MA (1964).

Boyer 73] R. S. Boyer and J. S. Moore, 'Proving Theories about LISP Functions,' pp. 486–493, in *Proc. 3rd International Joint Conference on Artificial Intelligence*, Stanford, CA (1973).

Brooks 85] R. A. Brooks, *Programming in Common-LISP*, John Wiley & Sons, NY (1985).

Burge 75] W. H. Burge, *Recursive Programming Techniques*, Addison-Wesley, Reading, MA (1975).

Chailloux 78] J. Chailloux, *VLISP 10.3, Manuel de Référence* (VLISP reference manual), RT–16–78, University of Paris 8–Vincennes (April 1978).

Chailloux 79] J. Chailloux, *VLISP8.2 Manuel de Reference* (VLISP reference manual), RT 11–79, Université Paris 8–Vincennes (August 1979).

Chailloux 79] J. Chailloux, *Le modèle VLISP: Description, Implementation et Evaluation* (The VLISP model: description, implementation and evaluation), Université Paris 7, LITP 80–20 (April 1980). thèse de 3ieme cycle.

Chailloux 84] J. Chailloux, LE_LISP de l'INRIA: *Le Manuel de référence* (INRIA's LE_LISP: reference manual), INRIA, Le Chesnay (December 1984).

Charniak 80] E. Charniak, C. K. Riesbeck, and C. V. McDermott, *Artificial Intelligence Programming*, Lawrence Erlbaum Associates, Hillsdale, NJ (1980).

Durieux 78] J. L. Durieux, 'T.LISP IRIS 80,' *LSI report no 130*, Universite Paul Sabatier-Toulouse (June 1978).

Farreny 84] H. Farreny, *Programmer en LISP* (Programming in LISP), Masson, Paris, New York (1984).

Friedman 74] D. Friedman, *The Little LISPer*, Science Research Associates, Palo Alto, CA (1974).

Gord 64] E. Gord, 'Notes on the Debugging of LISP Programs,' in *The Programming Language LISP: its Operation and Applications*, ed. E. C. Berkeley & D. G. Bobrow, MIT Press, Cambridge, MA (1964).

Greussay 76] P. Greussay, *VLISP: Structures et extensions d'un system LISP pour mini-ordinateur* (VLISP: structure and extension of LISP system for a mini-computer), RT 16–76, Département Informatique, Université Paris 8–Vincennes (January 1976).

Greussay 76a] P. Greussay, *Iterative Interprétations of Tail-Recursive LISP Procedures*, TR 20–76, Départment Informatique, Université Paris 8-Vincennes (September 1976).

[Greussay 77] P. Greussay, *Contribution a la définition interprétative et a l'implémentation de langages* (Contribution to the interpretative definition and the implementation of languages), Université Paris 7 (November 1977). Thèse d'Etat.

[Greussay 78] P. Greussay, *Le Système VLISP 16* (The VLISP 16 system), Ecole Polytechnique (December 1978).

[Greussay 79] P. Greussay, *VLISP–11: Manuel de Référence*, Départment Informatique Université Paris 8-Vincennes (1979).

[Greussay 79a] P. Greussay, 'Aides à la Programmation en LISP: outils d'observation et compréhension' (Aids to LISP programming: observation and comprehension tools), *Bullet du Groupe Programmation et Langages* (9), pp. 13–25 AFCET, Division Théorique Technique de l'informatique (October 1979).

[Greussay 82] P. Greussay, *Le Système VLISP-UNIX* (The VLISP-UNIX system), Département Informatique, Université Paris 8–Vincennes (February 1982). Manual.

[Henderson] P. Henderson, *Functional Programming Application and Implementation*, Universi of Newcastle upon Tyne, Prentice-Hall International.

[Knuth 68] D. E. Knuth, *The Art of Computer Programming, Vol. 1 Fundamental Algorithm* Addison-Wesley, Reading, MA (1968).

[Landin 64] J. Landin, 'The Mechanical Evaluation of Expressions', Vol. 6, *Computer Journ* (January 1964).

[Landin 65] P. J. Landin, "A Correspondence between ALGOL60 and Church's Lambd Notation", *Comm. ACM* 8(2–3), pp. 89–101 and 158–165 (1965).

[Laubsch 76] J. H. Laubsch, 'MACLISP Manual,' *CUU-Memo-3*, Universitat Stuttgar Stuttgart, West Germany (1976).

[Lombardi 64] L. A. Lombardi and B. Raphael, 'LISP a Language for an Increment Computer,' in, *The Programming Language LISP: its Operation and Applications*, ed. E. Berkeley & D. G. Bobrow, MIT Press, Cambridge, MA (1964).

[Maurer 73] W. D. Maurer, *A Programmer's Introduction to LISP*, American Elsevier, Ne York, NY (1973).

[McCarthy 62] J. McCarthy and C. Talcott, *LISP–Programming and Proving*, Stanford Universit Stanford, CA (1978) (Draft).

[Moon 74] D. A. Moon, 'MACLISP Reference Manual,' *MIT Project MAC*, Cambridge M (April 1974).

[Moon 79] D. A. Moon and D. Weinreb, *LISP Machine Manual*, MIT, Cambridge, M (January 1979).

[Perrot 79] J. F. Perrot, 'Sur la structure des interprètes LISP' (On the structures of LIS interpreters), in, *Proceedings of the 'Codage of Transductions' Colloquium*, Florence (17– Octobre 1979).

[Quam 72] L. H. Quam et W. Diffie, 'Stanford LISP 1.6 Manual', *SAILON* 28.6, Comput Science Dept., Stanford University (1972).

[Queinnec 82] C. Queinnec, *Langage d'un autre type: LISP* (Another type of language: LISP Eyrolles, Paris (1982).

[Queinnec 84] C. Queinnec, *LISP mode d'emploi* (LISP operating instructions), Eyrolles, Pa (1984).

[Ribbens 70] D. Ribbens, *Programmation Non Numerique: LISP 1.5* (Non-numerical program ming: LISP 1.5), Dunod, Paris (1970).

[Saint-James 84] E. Saint-James, 'Recursion is more Efficient than Iteration,' pp. 228–234, in *Proc. ACM LISP Conference 1984*, Austin, Texas (1984).

[Sandewall 75] E. Sandewall, 'Ideas about Management of LISP data bases,' *Al-Memo No 332*, MIT Artificial Intelligence Laboratory, Cambridge, MA (1975).

[Siklossy 76] L. Siklossy, *Let's Talk LISP*, Prentice Hall, Englewood Cliffs, NJ (1976).

[Steele 79] G. L. Steele and G. J. Sussman, 'Design of LISP-Based Processors,' *Memo 514, MIT Artificial Intelligence Laboratory* (March 1979).

[Steele 84] G. L. Steele Jr, *Common-LISP, the Language*, Digital Press, Burlington, Mass (1984).

[Stoyan 78] H. Stoyan, *LISP-Programmierhandbuch* (LISP programmer's manual), Akademie Verlag, Berlin, East Germany (1978).

[Stoyan 80] H. Stoyan, *LISP–Anwendungsgebiete, Grundbegriffe, Geschichte* (LISP–applications, basic concepts, history), Akademie Verlag, Berlin, East Germany (1980).

[Stoyan 84] H. Stoyan, *LISP Eine Einfuhrung in die Programmierung* (LISP, an introduction to programming), Springer Verlag, Berlin, Heidelberg, New York (1984).

[Teitelman 78] W. Teitelman, INTERLISP Reference Manual, Xerox, Palo Alto Research Center, Palo Alto (October 1978), 3rd revision.

[Weissman 67] C. Weissman, *LISP 1.5 Primer*, Dickenson Pub. Co., Belmont, CA (1967).

[Wertz 83] H. Wertz, 'An Integrated, Interactive and Incremental Programming Environment for the Development of Complex Systems,' pp. 235–250, in *Integrated Interactive Computing Systems*, ed. P. Degano & E. Sandewall, North-Holland, Amsterdam, New York, Oxford (1983).

[Wertz 87], *Automatic Correction and Improvement of Programs*, Ellis Horwood Ltd, Chichester, (1987).

[Wilensky 84] R. Wilensky, *Common LISPcraft*, Norton & Co, NY (1984).

[Winston 77] P. H. Winston, *Artificial Intelligence*, Addison Wesley (1977).

[Winston 81] P. H. Winston and B. K. P. Horn, *LISP*, Addison Wesley (1981).

21 STANDARD LISP FUNCTIONS

In this appendix we shall list all the standard LISP functions met in this book. For each function we give the page where it is defined, as well as its type, the number of arguments and a short description of its semantics. We shall distinguish between three types of functions: *SUBR*s, *MACRO*s and *FSUBR*s (in Le_LISP) or Special Forms (in Common-LISP). SUBR functions are those in which all the arguments are evaluated. MACROs are those functions which do not evaluate all their arguments and are written using the DEFMACRO function, i.e. which expand into other function calls. FSUBRs and Special Forms are functions with a variable number of arguments where the arguments are not necessarily evaluated.

The first part will give Le_LISP functions, the second part Common-LISP functions.

Le_LISP

"	character to delimit strings (page 68)
#'	short form for FUNCTION, used to quote functional arguments (page 103)
#\	printed representation for special characters (page 81)
#\BELL	the bell character, <ctrl> G
#\BS	the backspace character
#\CR	the 'carriage return' character
#\DEL	the delete character
#\ESC	the escape character
#\LF	the linefeed character
#\SP	the space character
#\TAB	the tabulation character
(,)	characters to delimit lists (page 1)
*	(* n_1 . . . n_N), SUBR with N arguments (page 40) computes the product of its arguments. If no argument is given * returns the value 1.
+	(+ n_1 . . . n_N), SUBR with N arguments (page 40) computes the sum of its arguments. If no argument is given, + returns the value 0.
,	inside a backquote, the comma indicates the value of an element (page 126)
,@	inside a *backquote*, the comma followed by an at-sign, indicates that the value of the following argument should be *spliced* in (page 126)

− — ($- n_1 \ldots n_N$), SUBR with N arguments (page 40)
computes the difference of its arguments. If no argument is given, − returns the value 0.

/ — ($/ n_1 \ldots n_N$), SUBR with N arguments (page 40)
successively divides the first argument by all others and returns the result.

1+ — (1+ n), SUBR with 1 argument (page 40)
computes the sum of n and 1.

1− — (1− n), SUBR with 1 argument (page 40)
computes the difference of n and 1.

; — character to begin comments. All characters following, until the end of line, are ignored

< — ($< n_1 \ldots n_N$), SUBR with n arguments (page 42)
if $n_1 < n_2 < \ldots < n_N$ then $<$ returns n_1, else NIL

<= — ($<= n_1 \ldots n_N$), SUBR with N arguments (page 43)
if $n_1 \leqslant n_2 \leqslant \ldots \leqslant n_N$ then $\leqslant$ returns n_1,
else NIL

= — ($= n_1 \ldots n_N$), SUBR with N arguments (page 42)
if $n_1 = n_2 = \ldots = n_N$ then $=$ returns n_1, else NIL

> — ($> n_1 \ldots n_N$), SUBR with N arguments (page 42)
if $n_1 > n_2 > \ldots n_N$ then $>$ returns n_1, else NIL

>= — ($>= n_1 \ldots n_N$), SUBR with N arguments (page 43)
if $n_1 \geqslant n_2 \geqslant \ldots \geqslant n_N$ then $\geqslant$ returns n_1,
else NIL

AND — (AND $s_1 \ldots s_N$), FSUBR with n arguments (page 74)
returns the value of s_N if the evaluation of all s_i returns a value different from NIL, otherwise it returns NIL

APPEND — (APPEND $l_1 \ldots l_N$), SUBR with N arguments (page 31)
returns the concatenation of the lists $l_1, l_2 \ldots l_N$.

APPLY — (APPLY fn $s_1 \ldots s_N$ l_1), SUBR with 2 or N arguments (page 102)
the arguments $s_1 \ldots s_N$ are optional. l_1 should be a list. It returns the value of the function fn applied to l_1, which is, if the arguments s_1 are present, augmented by these arguments.

ASSOC — (ASSOC s a_1), SUBR with 2 arguments (page 193)
s should be a key, a_1 should be an A-list. Returns the cons-cell of a_1 beginning with s.

ATOM — (ATOM s), SUBR with 1 argument (page 20)
returns T if s is an atom, otherwise it returns NIL.

CADDR — (CADDR l), SUBR with 1 argument (page 7)
returns the third element of a list

CADR — (CADR l), SUBR with 1 argument (page 7)
returns the second element of a list

CAR — (CAR l), SUBR with 1 argument (page 6)
returns the first element of a list

CDR — (CDR l), SUBR with 1 argument (page 6)
returns the list without its first element

CLOSE — (CLOSE *chan*), SUBR with 0 or 1 argument (page 136)
closes the channel *chan* and returns T. If no argument is supplied, CLOSE closes all channels

CONCAT — (CONCAT str_1 ... str_N), SUBR with N arguments (pages 80–81)
constructs a new symbol whose name is the concatenation of str_1, str_2, . . . str_N

COND — (COND l_1 . . . l_N), FSUBR (page 36)
COND evaluates the first element of each of the lists l_1, l_2, . . . l_i until the evaluation returns a value different from NIL, then, the remaining expressions of l_i are evaluated and the value of the last expression of l_i is returned. If all evaluations of the successive CARs return NIL, COND returns NIL

CONS — (CONS s_1 s_2), SUBR with 2 arguments (page 7)
constructs and returns a new list whose CAR is the value of s_1 and whose CDR is the value of s_2.

CONSP — (CONSP *s*), SUBR with 1 argument (page 21)
returns T if *s* is a list different from NIL

DE — (DE *symb lvar* s_1 . . . s_N), FSUBR (pages 15, 87–90)
permits us to construct a new user defined function of type EXPR. *symb* is the name of this new function. DE returns the name of the new function.

DECR — (DECR *symb n*), FSUBR (pages 173, 235)
decrements the value of *symb* and returns the new value. If a second argument is supplied, the *symb* is decremented *n* times.

DEFMACRO — (DEFMACRO *symb lvar* s_1 . . . s_N), FSUBR (page 167)
defines a new MACRO with the name *symb*. Returns *symb*.

DELETE — (DELETE *s l*), SUBR with 2 arguments (page 37)
returns the list *l* where all occurrences of *s* on the first level are deleted.

DF — (DF *symb lvar* s_1 . . . s_N), FSUBR (pages 90–91)
permits to construct a new user defined function of type FEXPR. *symb* is the name of this new function. DF returns the name of the new function.

DISPLACE — (DISPLACE l l_n), SUBR with 2 arguments (page 235)
replaces the CAR of the list l with the CAR of the list l_n, and replaces the CDR of the list l with the CDR of the list l_n.

DMC — (DMC *ch* () s_1 . . . s_N), FSUBR (page 122)
defines *ch* as a new macro-character. Returns *ch*.

DO — (DO *lv lr* s_1 . . . s_N), MACRO (page 184)
general iteration function: *lv* is a list of triplets: <variables, initial value, repetitition forms>.
lr is a list whose CAR is the iteration exit test and whose CDR is the list of expressions to evaluate when the test is non NIL. If the test is NIL the expressions s_1, s_2 . . . s_N are evaluated. Then the *repetition forms* are evaluated and assigned to the corresponding variables.

EQ — (EQ s_1 s_2), SUBR with 2 arguments (pages 34, 157–158) compares the addresses of s_1 and s_2. Returns T if they are identical, otherwise EQ returns NIL.

EQUAL — (EQUAL s_1 s_2), SUBR with 2 arguments (pages 34, 157–158) compares the structures s_1 and s_2. Returns T if they are similar, NIL otherwise.

EVAL — (EVAL *s*), SUBR with 1 argument (page 100) returns the value of *s*. This is the main function of the LISP interpreter.

EVENP — (EVENP *n*), SUBR with 1 argument (page 43) returns *n* if *n* is an even number, otherwise NIL.

EXPLODECH — (EXPLODECH *s*), SUBR with 1 argument (page 77) returns a list of all the characters composing the expression *s*

FORMAT — (FORMAT *dest ctrls* s_1 s_2 . . . s_N), SUBR with $_N$ arguments (pages 75–77)
FORMAT edits the string *ctrls* in the output channel *dest.ctrls* may contain directives. These are formed by the character ~ followed by a letter specifying the directive. Possible arguments are given as arguments 3 to N. Possible directives are:
~% insert an end of line
~A insert the argument as an ascii string
~D insert the argument as a decimal number
~O insert the argument as an octal number
~S insert the argument as an S-expression
~T insert a tabulation
~X insert the argument as a hexadecimal number
~~ insert a tilde

FUNCALL — (FUNCALL *fn* s_1 . . . s_N), SUBR with *N* arguments (page 177) returns the value of the application of the function *fn* to the arguments s_1, s_2 . . . s_N

GENSYM — (GENSYM), SUBR with 0 arguments (page 172) returns a new symbol of the style G*xxx*

GET — (GET *pl ind*), SUBR with 2 arguments (page 55) and
GETPROP — (GETPROP *pl ind*), SUBR with 2 arguments (page 55) returns the value associated to *ind* on the P-list of the symbol *pl*.

IF — (IF *t* s_1 s_2 . . . s_N), FSUBR (page 24) the first argument *t* is considered a test. If its evaluation yields a non-NIL value, IF returns the evaluations of s_1. If the evaluation of *t* yields NIL, the expressions s_2, s_3, . . . s_N are evaluated sequentially and the value of s_N is returned.

IMPLODECH — (IMPLODECH *s*), SUBR with 1 argument (page 79) *s* is a list of characters. IMPLODECH returns the expression as a result of giving these characters to the LISP reader.

INCHAN — (INCHAN *chan*), SUBR with 0 or 1 argument (page 136) associates the channel *chan* to the input stream. If no argument is supplied, INCHAN returns the current channel.

INCR — (INCR *symb n*), FSUBR (pages 173, 235)

increments the value of *symb* and returns the new value. If a second argument is supplied, the *symb* is incremented *n* times.

LAMBDA — (LAMBDA *lvar* $s_1 \ldots s_N$), FSUBR (page 180)
the value of a lambda expression is this lambda expression itself

LAST — (LAST s), SUBR with 1 argument (page 156)
returns the last cons cell of the list *s*. If *s* isn't a list, the argument *s* is returned.

LENGTH — (LENGTH *s*), SUBR with 1 argument (page 216)
returns the number of elements of the list *s*. If *s* isn't a list LENGTH returns 0

LET — (LET *lvar* $s_1 \ldots s_N$), FSUBR (page 45)
calls an anonymous user function of the EXPR type. *lvar* is a list of couples <variable, initial value>. All variables are bound to their initial value, then the expressions $s_1 \ldots s_N$ are evaluated

LETN — (LETN *symb lvar* $s_1 \ldots s_N$), FSUBR (page 47)
like LET but permits to give the temporary name *symb* to the function.

LIST — (LIST $s_1 \ldots s_N$), SUBR with N arguments (page 89)
returns the list composed of $s_1, s_2, \ldots s_N$

LOAD — (LOAD *file*), FSUBR (page 131)
loads the file *file* and returns the name of the file

LOADFILE — (LOADFILE *file i*), SUBR with 2 arguments (page 136)
like the previous function, except that 1) the arguments are evaluated, 2) the second argument *i* is an indicator telling the system if function re-definitions are allowed or not.

MACROEXPAND — (MACROEXPAND *s*), SUBR with 1 argument (page 172)
returns the expression *s* with all calls to macro-functions expanded. This is very useful for testing one's macro-functions.

MAPC — (MAPC *fn* $l_i \ldots l_N$), SUBR with N arguments (page 181)
applies the function *fn* to all the CARs of the lists $l_1 \ldots l_N$ as arguments. MAPC returns NIL

MAPCAR — (MAPCAR *fn* $l_1 \ldots l_N$), SUBR with N arguments (page 178)
like MAPC, but returns a list of the successive function applications.

MAX — (MAX $n_1 \ldots n_N$), SUBR with N arguments (page 228)
returns the largest value of $n_1, n_2, \ldots n_N$

MCONS — (MCONS $s_1 \ldots s_N$), SUBR with N arguments (page 170)
returns the list obtained by successively CONSing all the elements s_i.

MEMQ — (MEMQ *symb l*), SUBR with 2 arguments (page 213)
if *symb* is an element of the list *l*, MEMQ returns the sub-list beginning with *symb*, otherwise it returns NIL (the comparisons are done with the function EQ)

MEMBER — (MEMBER *s l*), SUBR with 2 arguments (page 213)
like MEMQ, but the comparisons are done with EQUAL.

MIN — (MIN $n_1 \ldots n_N$), SUBR with N arguments (page 228)
returns the smallest number of $n_1, n_2, \ldots n_N$

NCONC (NCONC l_1 . . . l_N), SUBR with N arguments (page 156)
physically concatenates all the lists l_i. Returns the new list l_1.

NCONC1 (NCONC1 l a), SUBR with 2 arguments (page 175)
physically adds to the list l the new last element a. Returns the new list l

NEQUAL (NEQUAL s_1 s_2), SUBR with 2 arguments (page 122)
this is the same as (NULL (EQUAL s_1 s_2))

NEXTL (NEXTL $symb_1$ $symb_2$), FSUBR (page 234)
$symb_1$ and $symb_2$ must be symbols. NEXTL returns the CAR of the value of $symb_1$ and assigns to $symb_1$ the CDR of its value. If $symb_2$ is present, it receives the value returned by NEXTL.

NTH (NTH n l), SUBR with 2 arguments (page 215)
returns the n-th element of l

NULL (NULL s), SUBR with 1 argument (page 22)
returns T if s evaluates to NIL, otherwise it returns NIL

NUMBERP (NUMBERP s), SUBR with 1 argument (page 22)
returns s if it is a number, otherwise it returns NIL

OBLIST (OBLIST), SUBR with 0 arguments (page 114)
returns the list of *all* known symbols

ODDP (ODDP n), SUBR with 1 argument (page 43)
returns n if it is an odd number, NIL otherwise

OPENI (OPENI *file*), SUBR with 1 argument (page 136)
opens the file *file* for input. Returns the channel associated to this file.

OPENO (OPENO *file*), SUBR with 1 argument (page 136)
opens the file *file* for output. Returns the channel associated to this file.

OR (OR s_1 . . . s_N), FSUBR (page 73)
evaluates one s_i after the other until a non-NIL value is obtained.

OUTCHAN (OUTCHAN *chan*), SUBR with 0 or 1 argument (page 136)
associates the channel *chan* to the current output stream, that is: all successive outputs will be on this channel. If no argument is supplied, OUTCHAN returns NIL

PEEKCH (PEEKCH), SUBR with 0 arguments (page 110)
returns the next character of the input stream *without* removing it.

PLENGTH (PLENGTH *str*), SUBR with 1 argument (page 78)
returns the length of the string *str*

PLIST (PLIST *pl* l), SUBR with 1 or 2 arguments (page 52)
if l is not supplied, PLIST returns the P-list of *pl*, otherwise the list l becomes the new P-list of *pl*

PLUS (PLUS n_1 . . . n_N), SUBR with N arguments (page 88)
returns the sum of n_1, n_2, . . . n_N. If no argument is supplied, PLUS returns 0

PRIN (PRIN s_1 . . . s_N), SUBR with N arguments (page 70)
prepares the expression s_1, s_2, . . . s_N for printing. Actual

	printing will only take place when the output buffer is emptied
PRINCH	(PRINCH *ch n*), SUBR with 1 or 2 arguments (page 72) edits *n* times the character *ch* in the output stream, returns *ch*
PRINT	(PRINT s_1 . . . s_N), SUBR with *N* arguments (page 63) edits and prints the expressions s_1, s_2, . . . s_N in the output stream
PROBEFILE	(PROBEFILE *file*), SUBR with 1 argument (page 136) returns T if the file *file* exists, NIL otherwise
PUTPROP	(PUTPROP *pl s ind*), SUBR with 3 arguments (page 54) associates *s* to *ind* on the P-list of *pl*. PUTPROP returns the value of *s*
QUO	(QUO n_1 n_2), SUBR with 2 arguments (page 41) returns the integer quotient of n_1/n_2
QUOTE	(QUOTE *s*), FSUBR (page 5) *s* is not evaluated and simply returned
READ	(READ), SUBR with 0 arguments (page 106) (reads and returns the next LISP expression from the current input stream
READCH	(READCH), SUBR with 0 arguments (page 110) reads and returns the next character in the current input stream
REM	(REM n_1 n_2), SUBR with 2 arguments (page 41) returns the remainder of the integer division of n_1/n_2
REMPROP	(REMPROP *pl ind*), SUBR with 2 arguments (page 56) removes from the P-list of *pl* the indicator *ind* and its associated value. It returns the previous P-list if the indicator *ind* exists, otherwise NIL
REVERSE	(REVERSE *l*), SUBR with 1 argument (page 28) returns the list l with its elements in reverse order
RPLACA	(RPLACA *l s*), SUBR with 2 arguments (page 152) physically replaces the CAR of *l* with the value of *s*. RPLACA returns the new list *l*
RPLACD	(RPLACD *l s*), SUBR with 2 arguments (page 152) physically replaces the CDR of the *l* with the value of *s*. RPLACD returns the new list *l*
SET	(SET *sym s*), SUBR with 2 arguments (page 162) changes the value of *sym* to *s*
SETQ	(SETQ sym_1 s_1 . . . , sym_N s_N), FSUBR (page 162) sym_1, . . ., sym_N are symbols (these arguments are not evaluated) s_1, . . . s_N are expressions. SETQ assigns to each sym_i the corresponding value of s_i. It returns the value of s_N
SQRT	(SQRT *n*), SUBR with 1 argument (page 179) returns the square root of *n*
STRING	(STRING *s*), SUBR with 1 argument (page 80) returns the string conversion of the argument
STRINGP	(STRINGP *s*), SUBR with 1 argument (page 69) returns T if *s* is a string, otherwise NIL
SYMEVAL	(SYMEVAL *sym*), SUBR with 1 argument (page 52)

returns the value of the symbol *sym*. This is the same as EVAL, but more efficient.

TERPRI — (TERPRI *n*), SUBR with 0 or 1 argument (page 70)
effects 1 or *n* newlines. Returns *n*

TIMES — (TIMES $n_1 \ldots n_N$), SUBR with *N* arguments (page 227)
returns the product of $n_1 * n_2 * \ldots * n_N$

TYI — (TYI), SUBR with 0 arguments (page 111)
reads a character on the keyboard and returns its internal code

TYO — (TYO $o_1 \ldots o_N$), SUBR with *N* arguments (page 111)
edits the characters of $o_1, \ldots o_N$ on the terminal

TYPECH — (TYPECH *s symb*), SUBR with 1 or 2 arguments (page 119)
returns the type of the character *c* if *symb* is not supplied, otherwise it changes the type of the character *c* to *symb*

UNLESS — (UNLESS *tst* $s_1 \ldots s_N$), FSUBR (page 61)
if the evaluation of *tst* is NIL, UNLESS evaluates in sequence $s_1, s_2, \ldots s_N$ and returns the value of s_N. Otherwise, if *tst* evaluates to a non-NIL value, UNLESS returns NIL

UNTIL — (UNTIL *tst* $s_1 \ldots s_N$), FSUBR (page 183)
while the evaluation of *tst* is NIL, UNTIL evaluates in sequence $s_1, \ldots s_N$. It returns the first value of tst different from NIL

WHEN — (WHEN *tst* $s_1 \ldots s_N$), FSUBR (page 61)
if the evaluation of *tst* is non-NIL, WHEN evaluates in sequence $s_1, s_2, \ldots s_N$ and returns the value of s_N. Otherwise, if *tst* evaluates to a NIL value, WHEN returns NIL

WHILE — (WHILE *tst* $s_1 \ldots s_N$), FSUBR (page 183)
while the evaluation of *tst* is non-NIL, WHILE evaluates in sequence $s_1, \ldots s_N$. It returns NIL when the evaluation of *tst* is NIL

ZEROP — (ZEROP *n*), SUBR with 1 argument (page 42)
returns T if *n* equals 0, otherwise NIL

` — backquote macro-character (page 126)

| — multiple escape character (page 118)

' — the QUOTE macro-character (page 5)

COMMON-LISP

" — character to delimit strings (page 68)

#' — short form for FUNCTION, used to quote functional arguments (page 103)

#\ — printed representation for special characters (page 81)

#\Newline — the 'newline' character (page 113)

#\Page — the 'page' character (page 113)

#\Return — the 'carriage return' character (page 113)

#\Space — the 'white space' character (page 113)

#\Tab — the tabulation character (page 113)

&KEY — the 'keyword' parameter keyword (page 97)

&OPTIONAL — the 'optional' parameter keyword (page 95)

&REST — the 'rest' parameter keyword (page 92)

&WHOLE — the 'all' parameter keyword (page 184)

(,) — characters to delimit lists (page 1)

***** — (* $n_1 \ldots n_N$), SUBR with N arguments (page 40)
computes the product of its arguments. If no argument is given, * returns the value 1.

+ — (+ $n_1 \ldots n_N$), SUBR with N arguments (page 40)
computes the sum of its arguments. If no argument is given, + returns the value 0.

, — inside a backquote, the comma indicates the value of an element (page 126)

,@ — inside a *backquote*, the comma followed by an at-sign, indicates that the value of the following argument should be *spliced* in. (page 126)

− — (− $n_1 \ldots n_N$), SUBR with N arguments (page 40)
computes the difference of its arguments. If only one argument is given, − returns the negative of that argument.

/ — (/ $n_1 \ldots n_N$), SUBR with N arguments (page 40)
successively divides the first argument by all the others and returns the result. If only one argument is given, / reciprocates its argument. Note the / produces a ratio if the mathematical quotient of two integers is not an exact integer.

1+ — (1+ n), SUBR with 1 argument (page 40)
computes the sum of n and 1.

1− — (1− n), SUBR with 1 argument (page 40)
computes the difference of n and 1.

: — indicator of a constant. Mainly used in keyword parameters

:DIRECTION — keyword for input/output operations (pages 132–133)

:INPUT — keyword for input operations (pages 132–133)

:IO — keyword for input/output operations (pages 132–133)

:OUTPUT — keyword for output operations (pages 132–133)

:PROBE — keyword for testing the existence of files (pages 132–133)

; — character to begin comments. All characters following, until the end of line, are ignored

< — ($< n_1 \ldots n_N$), SUBR with N arguments (page 42)
if $n_1 < n_2 < \ldots < n_N$ then < returns T, else NIL

<= — ($<= n_1 \ldots n_N$), SUBR with N arguments (page 43)
if $n_1 \leqslant n_2 \leqslant \ldots \leqslant n_N$ then ≤ returns T, else NIL

= — ($= n_1 \ldots n_N$), SUBR with N arguments (page 42)
if $n_1 = n_2 = \ldots = n_N$ then = returns T, else NIL

> — ($> n_1 \ldots n_N$), SUBR with N arguments (page 42)
if $n_1 > n_2 > \ldots > n_N$ then < returns T, else NIL

>= — ($>= n_1 \ldots n_N$). SUBR with N arguments (page 43)
if $n_1 >= n_2 \geqslant \ldots \geqslant n_N$ then = returns T, else NIL.

AND — (AND $s_1 \ldots s_N$), MACRO with n arguments (page 74)
returns the value of s_N if the evaluation of all s_i returns a value different from NIL, otherwise it returns NIL

APPEND — (APPEND l_1 . . . l_N), SUBR with N arguments (page 31)
returns the concatenation of the lists l_1, l_2 . . . l_N.

APPLY — (APPLY *fn* s_1 . . . s_N l_l), SUBR with 2 or n arguments (page 102)
the arguments s_1 . . . s_N are optional. l_1 should be a list. It returns the value of the function *fn* applied to l_1, which is, if the arguments s_i are present, augmented by these arguments.

ASSOC — (ASSOC s a_1), SUBR with 2 arguments (page 193)
s should be a key, a_1 should be an A-list. Returns the cons-cell of a_1 beginning with s.

ATOM — (ATOM s), SUBR with 1 argument (page 20)
returns T if s is an atom, otherwise it returns NIL.

CADDR — (CADDR l), SUBR with 1 argument (page 7)
returns the third element of a list

CADR — (CADR l), SUBR with 1 argument (page 7)
returns the second element of a list

CAR — (CAR l), SUBR with 1 argument (page 6)
returns the first element of a list

CDR — (CDR l), SUBR with 1 argument (page 6)
returns the list without its first element

CHAR> — (CHAR> c_1 . . . c_N), SUBR with N arguments (page 176)
returns T if $c_1 > c_2 > . . . > c_N$, otherwise NIL

CHARACTERP — (CHARACTERP s), SUBR with 1 argument (page 129)
returns T if s is a character object, otherwise NIL

CLOSE — (CLOSE *stream*), SUBR with 1 argument (page 132)
closes the *stream* and returns T

CONCATENATE — (CONCATENATE *result-type* seq_1 . . . seq_N) SUBR with N arguments (pages 80–81)
the result is a concatenation of all the sequences seq_1 to seq_N. *result-type* specifies the type of the result which must be a subtype of sequence.

COND — (COND l_1 . . . l_N), MACRO (page 36)
COND evaluates the first element of each of the lists l_1, l_2, . . . l_i until the evaluation returns a value different from NIL, then, the remaining expressions of l_i are evaluated and the value of the last expression of l_i is returned. If all evaluations of the successive CARs return NIL, COND returns NIL.

CONS — (CONS s_1 s_2), SUBR with 2 arguments (page 7)
constructs and returns a new list whose CAR is the value of s_1 and whose CDR is the value of s_2.

CONSP — (CONSP s), SUBR with 1 argument (page 21)
returns T if s is a list different from NIL

DECLARE — (DECLARE (*typespec* var_1 . . . var_N)) special form (page 191)
This is a special form permitting to imbed declarations within executable code. The only *type specifier* we have met in this book is the type SPECIAL, which declares the variables var_1, . . ., var_N to be dynamically scoped.

DEFMACRO (DEFMACRO *symb lvar* $s_1 \ldots s_N$), FSUBR (page 167)
defines a new MACRO with the name *symb*. Returns *symb*.

DEFUN (DEFUN *symb layer* $s_1 \ldots s_N$), MACRO (pages 15, 92–98)
permits to construct a new user defined function. *symb* is the name of this new function. DEFUN returns the name of the new function.

DELETE (DELETE *s l*), SUBR with 2 arguments (pages 37, 79–80)
returns the list *l* where all occurrences of *s* on the first level deleted.

DO (DO *lv lr* $s_1 \ldots s_N$), MACRO (page 184)
general iteration function: *lv* is a list of triplets: <variables, initial value, repetition forms>. *lr* is a list whose CAR is the iteration exit test and whose CDR is the list of expressions to evaluate when the test is non NIL. If the test is NIL the expressions $s_1, s_2 \ldots s_N$ are evaluated. Then the *repetition forms* are evaluated and assigned to the corresponding variables.

ELT (ELT *seq n*), SUBR with 2 arguments (page 81)
returns the *n*th element of sequence *seq*. Note that the first element of a sequence has index 0.

EQ (EQ s_1 s_2), SUBR with 2 arguments (pages 34, 157–158)
compares the addresses of s_1 and s_2. Returns T if they are idential, otherwise EQ returns NIL

EQUAL (EQUAL s_1 s_2), SUBR with 2 arguments (pages 34, 157–158)
compares the structures s_1 and s_2. Returns T if they are similar, NIL otherwise.

EVAL (EVAL *s*), SUBR with 1 argument (page 100)
returns the value of *s*. This is the main function of the LISP interpreter.

EVENP (EVENP *n*), SUBR with 1 argument (page 43)
returns T if *n* is an even number, otherwise NIL

FLOOR (FLOOR n_1 n_2), SUBR with 1 or 2 arguments (pages 41, 49)
if only one argument is present, FLOOR returns the largest integer equal or smaller than n_1. With two arguments, FLOOR returns the largest integer equal or smaller than the quotient of n_1/n_2

FORMAT FORMAT *dest ctrls* s_1 $s_2 \ldots s_N$), SUBR with N arguments (page 75–77)
FORMAT edits the string *ctrls* in the output channel *dest*. *ctrls* may contain directives. These are formed by the character ~ followed by a letter specifying the directive. Possible arguments are given as arguments 3 to N. Possible directives are:

~% insert an end of line
~A insert the argument as an ascii string
~D insert the argument as a decimal number
~O insert the argument as an octal number
~S insert the argument as an S-expression

~T insert a tabulation
~X insert the argument as a hexadecimal number
~~ insert a tilde

FUNCALL
(FUNCALL *fn* $s_1 \ldots s_N$), SUBR with N arguments (page 177)
returns the value of the application of the function *fi* to the arguments $s_1, s_2 \ldots s_N$

GENSYM
(GENSYM *s*), SUBR with 0 or 1 argument (page 172)
if no argument is present, returns a new symbol of the style G*xxx*. If the argument is a number, then the symbol returned is G*number*, if it is a string the symbol returned is *symbol* followed by a number.

GET
(GET *pl ind def*), SUBR with 2 or 3 arguments (page 55)
without the optional third argument, GET returns the value associated to *ind* on the P-list of the symbol *pl*. If the third argument is present, it gives the default value for NIL. This permits us to distinguish between NIL as a value associated to *ind* or as the value of GET if *ind* doesn't exist on *pl*.

GET-MACRO-CHARACTER
(GET-MACRO-CHARACTER *c*), SUBR with 1 argument (page 127)
returns the function associated with the character *c*. If *c* does not have an associated function, it returns NIL

IF
(IF *t then else*), Special form (page 24)
the first argument *t* is considered a test. If its evaluation yields a non-NIL value, IF returns the evaluations of *then*. If the evaluation of *t* yields NIL, the value of *else* is returned.

LABELS
(LABELS *sym lvar* $s_1 \ldots s_N$), Special form (page 46)
permits the local definition of a function with name *sym*. *lvar* is a list of variables and $s_1, \ldots, s_N$ is the body of the function.

LAST
(LAST), SUBR with 1 argument (page 156)
returns the last cons cell of the list *l*.

LENGTH
(LENGTH *s*), SUBR with 1 argument (pages 79–80, 216)
returns the number of elements of the sequence *s*.

LET
(LET *lvar* $s_1 \ldots s_N$), Special Form (page 45)
calls an anonymous user function of the EXPR type. *lvar* is a list of couples <variable, initial value>. All variables are bound to their initial value, then the expressions $s_1 \ldots s_N$ are evaluated.

LIST
(LIST $s_1 \ldots s_N$), SUBR with N arguments (page 89)
returns the list composed of $s_1, s_2, \ldots s_N$

LOAD
(LOAD *file*), FSUBR (page 131)
loads the file *file* and returns the name of the file

MACROEXPAND
(MACROEXPAND *s*), SUBR with 1 argument (page 172)
returns the expression *s* with all calls to macro-functions expanded. This is very useful for testing one's macro-functions.

MAPC
(MAPC *fn* $l_1 \ldots l_N$), SUBR with N arguments (page 181)
applies the function *fn* to all the CARs of the lists $l_1 \ldots l_N$ as arguments. MAPC returns NIL

MAPCAR — (MAPCAR *fn* l_1 . . . l_N), SUBR with N arguments (page 178)
like MAPC, but returns a list of the successive function applications.

MAX — (MAX n_1 . . . n_N), SUBR with N arguments (page 228)
returns the largest value of n_1, n_2, . . . n_N

MEMBER — (MEMBER *s l*), SUBR with 2 arguments (page 213)
The list *l* is searched for an occurrence of *s*. If it is found, then the tail of the list *l* beginning with *s* is returned, if none is found, MEMBER returns NIL

MIN — (MIN n_1 . . . n_N), SUBR with N arguments (page 228)
returns the smallest number of n_1, n_2, . . . n_N

NCONC — (NCONC l_1 . . . l_N), SUBR with N arguments (page 156)
physically concatenates all the lists l_i. Returns the new list l_1.

NTH — (NTH *n l*), SUBR with 2 arguments (page 215)
returns the *n*-th element of *l*. Note that the CAR of the list is the 'zeroth' element.

NULL — (NULL *s*), SUBR with 1 argument (page 22)
returns T if *s* evaluates to NIL, otherwise it returns NIL

NUMBERP — (NUMBERP *s*), SUBR with 1 argument (page 22)
returns *s* if it is a number, otherwise it returns NIL

ODDP — (ODDP *n*), SUBR with 1 argument (page 43)
returns *n* if it is an odd number, NIL otherwise

OPEN — (OPEN *file* :DIRECTION *dir*), SUBR with 2 arguments (pages 132–133)
returns a stream connected to the file *file*. Four directions are possible: :INPUT, an input stream is returned, :OUTPUT an output stream is returned, :IO, the result will be a bidirectional stream, and :PROBE, the result will be a no-directional stream if the file *file* exists, or NIL if the file *file* doesn't exist.

OR — (OR s_1 . . . s_N), MACRO (page 73)
evaluates one s_i after the other until a non-NIL value is obtained.

PEEK-CHAR — (PEEK-CHAR *peek-type input-stream eof-error eof-val rec-p*) SUBR with 0 to 5 arguments (page 110)
without arguments PEEK-CHAR returns the next character of the input stream *without* removing it. If *peek-type* is T, PEEK-CHAR skips over white space characters before performing the peeking operation. If *peek-type* is a character, then PEEK-CHAR skips over input characters until one which is equal is found.
If *input-stream* is non-NIL, then the peeking operation takes place in the given input stream.
If *eof-error* is T (the default), then reading past an end of file signals an error, otherwise no error is signalled and *eof-val* is returned.
If *rec-p* is non-NIL, then this specifies that the call to PEEK-CHAR is embedded in call to READ. This is the case if it is used inside macro-character definitions.

PRIN1 (PRIN1 *s stream*), SUBR with 1 or 2 arguments (page 70)
prints *s* on *stream* (which defaults to the standard output). PRIN1 outputs escape characters and double quotes so that the output is LISP-readable.

PRINC (PRINC *s stream*), SUBR with 1 or 2 arguments (page 72)
prints *s* on *stream* (which defaults to the standard output). No escape characters and double quotes are output, so the output is human-readable.

PRINT (PRINT *s stream*), SUBR with 1 or 2 arguments (page 63)
this is like PRIN1, except that the output is preceded by a newline.

QUOTE (QUOTE *s*), MACRO (page 5)
s is not evaluated and simply returned

READ (READ *input-stream eof-error eof-val rec-p*) SUBR with 0 to 4 arguments (pages 106, 134)
returns the next LISP expression read from *input-stream* (which defaults to the standard input;) T is considered terminal input/output). The other arguments have the same meaning as in PEEK-CHAR.

READ-CHAR (READ-CHAR *input-stream eof-err eof-val rec-p*) SUBR with 0 to 4 arguments (page 110)
like PEEK-CHAR except that the next character is taken off the *input-stream* and no longer available.

READ-CHAR-NO-HANG (READ-CHAR-NO-HANG *input-stream eof-error eof-val rec-p*) SUBR with 0 to 4 arguments (page 110)
like READ-CHAR, but if no character is available, these functions return immediate NIL and doesn't wait for a character.

REDUCE (REDUCE *fn seq* :FROM-END :START :END :INITIAL-VALUE) SUBR with 2 to 6 arguments (page 81–82)
This function combines all elements of a sequence using a binary operation *fn*. This reduction is left-associative, unless the :FROM-END argument is T (default is NIL) in which case it is right-associative. If an :INITIAL-VALUE is given it is placed before the sequence (after it if :FROM-END is T) and included in the reduction operation. The keywords :START and :END may specify a subsequence of *seq*, if this is done, then the reduction applies only to this subsequence

REM (REM n_1 n_2), SUBR with 2 arguments (page 41)
returns the remainder of the integer division of n_1/n_2

REMPROP (REMPROP *pl ind*), SUBR with 2 arguments (page 56)
removes from the P-list of *pl* the indicator *ind* and its associated value. It returns the previous P-list if the indicator *ind* exists, otherwise NIL

REVERSE (REVERSE *seq*), SUBR with 1 argument (pages 28, 79–80)
returns the sequence *seq* with its element in reverse order

RPLACA (RPLACA *l s*), SUBR with 2 arguments (page 152)

physically replaces the CAR of *l* with the value of *s*. RPLACA returns the new list *l*

RPLACD — (RPLACD *l s*), SUBR with 2 arguments (page 152)
physically replaces the CDR of the *l* with the value of *s*. RPLACD returns the new list *l*

SET — (SET *sym s*), SUBR with 2 arguments (page 162)
changes the value of sym to *s*

SET-MACRO-CHARACTER — (SET-MACRO-CHARACTER *c fn ntp*) SUBR with 2 or arguments (page 123)
SET-MACRO-CHARACTER causes the character *c* to be macro-character with the associated function *fn*. If *ntp* is non NIL (it defaults to NIL), then it will be a non terminating macro character which may be embedded within extended tokens. SET MACRO-CHARACTER returns T.

SET-SYNTAX-FROM-CHAR — (SET-SYNTAX-FROM-CHAR c_1 c_2), SUBR with 2 argument (page 121)
This makes the syntax of the character c_1 be the same as th syntax of the character c_2.

SETF — (SETF pl_1 v_1 . . . pl_N v_N), MACRO (pages 55, 164)
this is the generalized assignment function: it takes the place pl_1, pl_2, . . ., pl_N and assigns them the new values v_1, v_2, . . . v_N. The places pl_i may be any variable, or any place compute with function calls using any of the following known LISP functions: CAR, CDR and their combinations, NTH, GET SYMBOL-VALUE, SYMBOL-FUNCTION, SYMBOL-PLIST.

SETQ — (SETQ sym_1 s_1 . . . sym_N s_N), MACRO (page 162)
sym_1, . . ., sym_N are symbols (these arguments are not evaluated s_1, . . ., s_N are expressions. SETQ assigns to each sym_i th corresponding value of s_i. It returns the value of s_N

SQRT — (SQRT *n*), SUBR with 1 argument (page 179)
returns the square root of *n*

STRING — (STRING *s*), SUBR with 1 argument (page 80)
returns the string conversion of the argument

STRINGP — (STRINGP *s*), SUBR with 1 argument (page 69)
returns T if *s* is a string, otherwise NIL

SYMBOL-PLIST — (SYMBOL-PLIST *sym*), SUBR with 1 argument (page 52)
returns the P-list associated to the symbol *sym*

SYMBOL-VALUE — (SYMBOL-VALUE *sym*), SUBR with 1 argument (page 52)
returns the current value of the dynamic (special) symbol *sym*

TERPRI — (TERPRI *stream*), SUBR with 0 or 1 argument (page 70)
effects a newlines on the stream *stream* or on the standard outpu if no stream is specified. Returns NIL

UNLESS — (UNLESS *tst* s_1 . . . s_N), MACRO (page 61)
if the evaluation of *tst* is NIL, UNLESS evaluates in sequenc s_1, s_2, . . . s_N and returns the value of s_N. Otherwise, if *t* evaluates to a non-NIL value, UNLESS returns NIL

WHEN — (WHEN *tst* s_1 . . . s_N), MACRO (page 61)

if the evaluation of *tst* is non-NIL, WHEN evaluates in sequence s_1, s_2, . . . s_N and returns the values of s_N. Otherwise, if *tst* evaluates to a NIL value, WHEN returns NIL

WRITE-CHAR — (WRITE-CHAR *c stream*), SUBR with 1 or 2 arguments (page 111)
outputs the character *c* to the stream *stream* or to standard output if *stream* is not specified. WRITE-CHAR returns *c*.

ZEROP — (ZEROP *n*), SUBR with 1 argument (page 42)
returns T if *n* equals 0, otherwise NIL

backquote macro-character (page 126)

multiple escape character (page 118)

the QUOTE macro-character (page 5)

INDEX